AF539518

Molecular, Enzymatic and Biochemical Assays for Agricultural Pests

NIPA® GENX ELECTRONIC RESOURCES & SOLUTIONS P. LTD.
New Delhi-110 034

About the Editors

Dr. Srinivasa N. Ph.D. in Entomology from the Indian Agricultural Research Institute, New Delhi (India). He is working as an assistant professor of entomology at Banaras Hindu University. Dr. Srinivasa N area of specialization is Insect Molecular Biology, Insect Ecology and Integrated Pest Management. He has more than four years' experience in teaching and research. He has supervised 2 Ph.D. and 6 M.Sc. students. He has published 16 research publications in journals and 5 book chapters. Dr. Srinivasa N. is a life fellow of the Entomological Society of India and a reviewer of many entomology journals of international repute.

Dr. Bhupendra Kumar, Ph.D. in Zoology from University of Lucknow (India) is presently working as an Assistant Professor in Department of Zoology, Banaras Hindu University, Varanasi (India). Having specialization in Entomology, Dr. Kumar is working in the area of insect ecology, insect molecular science, insect behaviour and management of insect pests and weeds. He has a teaching experience of 9 years. He has supervised 01-Ph.D. and 21-M.Sc. dissertation student. 04-Ph.D. scholars are still working under his supervision. He has published 43-research articles, 10-book chapters and 2-books in his academic career. Dr. Kumar is the Fellow of the Royal Entomological Society, United Kingdom and the Life-Fellow of Entomological Society of India, New Delhi. He is also the reviewer of many entomological journals of international repute.

Dr. Rakesh Verma is currently working as Assistant Professor in the department of Zoology, Banaras Hindu University. He works in the area of molecular endocrinology and mammalian reproductive physiology along with impacts of various agents causing detrimental effect on male/female reproductive health, immune status and metabolic physiology. He has completed his B.Sc. (Hons) Zoology, M.Sc. Zoology and PhD (Zoology) from Department of Zoology, Banaras Hindu University, Varanasi. He has been awarded BHU Gold medal and numerous other awards for first rank in M.Sc. and B.Sc. He is engaged in teaching and research since October, 2015 upon his appointment in Department of Zoology, BHU. He has experience of successful supervision of PhD and postdoctoral students. He has guided numerous M.Sc. dissertations/project works. He has published numerous papers, book chapters in international/national journals and publishers. He has experience of handling various research projects. He has also served as reviewer of various international journals, participated in and organised various national and international seminars/ workshops and symposium.

Molecular, Enzymatic and Biochemical Assays for Agricultural Pests

Editors

Srinivasa N
Assistant Professor
Department of Entomology and Agricultural Zoology

Bhupendra Kumar
Assistant Professor
Department Zoology

Rakesh Verma
Assistant Professor
Department Zoology

Banaras Hindu University, Varanasi-221005, Uttar Pradesh, India

NIPA® GENX ELECTRONIC RESOURCES & SOLUTIONS P. LTD.
New Delhi-110 034

NIPA® GENX ELECTRONIC RESOURCES & SOLUTIONS P. LTD.

101,103, Vikas Surya Plaza, CU Block
L.S.C.Market, Pitam Pura, New Delhi-110 034
Ph : +91 11 27341616, 27341717, 27341718
E-mail: newindiapublishingagency@gmail.com
www: www.nipabooks.com

For customer assistance, please contact
Phone: + 91-11-27 34 17 17
Fax: + 91-11- 27 34 16 16

Print ISBN: 978-93-58873-13-9

ebook ISBN: 978-93-58877-22-9

Composed and Designed by NIPA®.

AN INSTITUTION OF NATIONAL IMPORTANCE ESTABLISHED BY AN ACT OF PARLIAMENT

An Institution devoted to nation and character building since 1916

निदेशक कार्यालय
कृषि विज्ञान संस्थान
Office of the Director
Institute of Agricultural Sciences

Prof. S.V.S. Raju
Director,
Institute of Agricultural Sciences

Foreword

Agricultural pests, including insects, pathogens, weeds, vertebrate pests, etc., cause massive damage to crops (average 15%). Identifying these pests is the first challenging task in taking timely management measures. Due to the declining number of taxonomists, identifying pests has become challenging. Traditional methods of pest identification based on morphological techniques are error-prone, time-consuming, and require a specialized taxonomist to identify them. Molecular tools offer unparalleled precision and accuracy in identifying and characterizing agricultural pests. Unlike traditional methods that rely on visual inspection or symptom-based diagnosis, molecular techniques such as DNA barcoding and species-specific markers enable the precise identification of pests at various life stages, including eggs, larvae, and adults. DNA barcoding is a method of specimen identification using short, standardized segments of DNA. The development of a DNA barcode unique to each species can be used for identification by comparing the reference library. Species-specific markers have become an invaluable tool to quickly identify the species irrespective of sex and stage of the organism, for example, insects. These markers will reduce the cost of sequencing and help in the timely, precise identification of the pest. Real-time PCR is a variation of the PCR assay to allow monitoring of the PCR progress in real time. Current applications of real-time PCR include gene expression analysis for various purposes, detection and quantification of pathogens, detection of genetically modified organisms, detection of allergens, species identification, etc. Further, antioxidant enzymes scavenge the free radicals that are generated during the various biological processes.

The workshop, entitled "National Workshop and Hands-on Training on Molecular Tools and Diagnostics of Agricultural Pests," commencing from 8th-10th April, 2024 aims to provide participants with a comprehensive understanding of the latest molecular techniques and diagnostic tools available for the identification, monitoring, and management of agricultural pests. I congratulate the organizers for this workshop and hope the participants will be enriched with molecular techniques for detection and diagnosis of agricultural pests.

Prof. S.V.S. Raju
Director

T &F: Director : 91-542-2368993, 6702571
Email: director.ias.bhu@gmail.com
W : www.bhu.ac.in

Preface

Identification of pests in agricultural and horticultural crops is the first step in the decision-making process for pest management. The classical approach to identification is based on morphological techniques, which have their own limitations, like sexual dimorphism and the loss of characters in a preserved specimen over a period, and require specialized taxonomists. DNA barcoding is one molecular approach that helps in the identification of pests by sequencing the barcode gene and comparing the barcode gene sequences in reference databases such as NCBI and BOLD. This method is available for plants, animals, microbes, etc. However, for DNA barcoding, one has to know basic molecular techniques such as DNA extraction, polymerase chain reaction (PCR), gel electrophoresis, primer design, and basic bio-informatic analysis. This book covers all those basic techniques required for DNA barcoding pests. Further, the book deals with protein estimation by Broadford and the immunoblotting technique, which deals with protein separation and identifying specific proteins in a tissue homogenate. Real-time polymerase chain reaction, also known as quantitative PCR, is a modification of the PCR strategy that allows real-time visualization of PCR progress. The real-time PCR technique has many applications, such as pest and disease diagnostics, gene expression analysis, bioremediation monitoring, and genetically modified organism detection and quantification. A chapter in the book deals with primer design for RT-PCR and data analysis.

Biochemicals such as proteins, carbohydrates, and lipids are essential for the survival and fitness of all organisms, including insects. However, their content will vary when they are exposed to stress conditions such as high temperatures, nutrient-deficient food, etc. To quantify the same, the protocols for estimation of glucose, soluble proteins, and triglycerides are presented in a chapter. Further, one of the chapters deals with procedures for the estimation of enzymes such as superoxide dismutase (SOD), catalase (CAT), and lactoperoxidase. From this, one can gain insight into an organism's ability to combat oxidative stress. The book "Molecular, Enzymatic and Biochemical Assays for Agricultural Pests" deals with basic molecular techniques from DNA extraction to Sequence submission in NCBI database. Further on protein estimation, immunoblotting and real-time PCR. Followed by biochemical assays such as glucose, protein triglycerides estimation along with stress responsive enzymes. It is anticipated that readers from a variety of fields will find the techniques discussed here useful for their own research.

Srinivasa N
Bhupendra Kumar
Rakesh Verma

Contents

1

DNA Extractions from Insects

Twinkle Sinha, Varun Arya and Srinivasa Narayana

Department of Entomology & Agricultural Zoology, Institute of Agricultural Sciences, Banaras Hindu University, Varanasi - 221005, Uttar Pradesh, India

Protocol for DNA extraction using the Qiagen DNeasy Blood and Tissue Kit (Catalogue No. 69504) presented here

Contents of the kit: DNeasy mini spin columns (2 ml), collecting tubes (2 ml), tissue lysis buffer (ATL), lysis buffer (AL), wash buffer 1 (AW1), wash buffer 2 (AW2), elution buffer (AE) and proteinase-K.

Note: Check for the validity of the product before dilution. Dilute the ATL and AL buffer with molecular gradient ethyl alcohol to a specified volume mentioned on the bottle before starting experiments. Keep a tick mark the bottle cap after dilution with ethyl alcohol for easy identification.

Equipments required: Microscope, centrifuge, vortex mixture, water bath, pestle and mortar/homogeniser, micropipettes, micro tips, eppendorf tubes, needle, micro scissor, forceps, camel hair brush, thermometer, tube stand, masks and gloves.

Procedure:

- Put on a fresh pair of gloves, mask and clean lab coat (if available).
- Clean the working area, pipettes, tip boxes, tube bottles and tube stand with 70% ethyl alcohol at the beginning.
- First observe the specimen under microscope for any external infection/ parasitism and remove the dust with 90% ethyl alcohol (hereinafter alcohol), prior to dissection.
- Dissect the specimen with the help of sterilized micro scissors/needle/ forceps. (Dissection is to remove specified part of insect such as leg, wing, antennae, body part etc.)
- Dry the specimen on clean tissue paper to let the alcohol evaporate.

- Switch on water-bath to maintain the temperature at 56ºC.
- Label the 1.5 ml autoclaved eppendorf tubes, gently transfer the sample (Insect) in the tubes with camel hair brush and add 30 µl of the ATL buffer.

  ***Note:** The quantity of the ATL buffer can increased/decreased based on the size of the specimen. For example, to extract DNA from the brown planthopper, we need to add 30 µl of the ATL buffer initially, followed by adding the buffer three times (50 µl each) by washing pestle, after homogenization.

- Crush the samples with pestle or homogeniser till disintegration. (This is the most crucial step as it enhances the quality and quantity of DNA), followed by addition of 50 µl of ATL buffer three times, as mentioned above.
- Discard the pestle (Reuse the pestle after washing and autoclave) and add 20 µl of proteinase-K followed by mixing with vortex.
- Keep the samples in hot water bath at 56°C for 2-3 hours

(The duration of incubation in hot water bath varies with samples to sample. For example shorter incubation period for soft bodied insects and longer duration for hard bodied insects).

- Take the samples out and add 200 µl AL buffer followed by immediate mixing by vortex. Further add 200 µl of ethyl alcohol (100%) and mix immediately by vortex.
- Spin the centrifuge tubes in mini-centrifuge to settle down the unwanted particles (spin slightly).
- Label the DNeasy mini spin column (attached with a 2 ml) and transfer the supernatant in it (approx. 600 µl).
- Centrifuge it at 8000 revolutions per minute (rpm) for 1 minute and discard the collection tube.
- Place the spin column in a new 2 ml collection tube and add 500 µl AW1 buffer, centrifuge at 8000 rpm for 1 minute and discard the collecting tube.
- Place the spin column in a 2 ml collecting tube, add 500 µl of the AW2 buffer and centrifuge it at 14000 rpm for 3 minutes.
- Dispose the supernatant liquid from the collecting tube and dry spin without adding anything at 14000 rpm for 1 minute.

- Discard collection tubes and transfer the spin columns to properly labelled eppendorf tubes (1.5 ml) and keep it open at room temperature for 3-5 minutes for proper evaporation of the alcohol.
- Elute the DNA by adding 30 µl of AE buffer (add 15 µl two times) to the centre of spin column in the tube and leave it for 2 minutes (Note: the amount of AE buffer added will depend upon the size of sample).
- Centrifuge the tubes at 14000 rpm for 1 minute at room temperature and discard the spin column.

DNA will be collected in 1.5 ml eppendorf tube. Transfer it to -20°C for long storage.

***Note:** The experiment must be conducted in clean area properly wiped with 70% ethyl alcohol. Gloves, mask and lab coat must be worn throughout the experiment. Crude genomic DNA can be checked for quality and quantity similar to PCR product through agarose gel electrophores.

Acknowledgement

Dr. Srinivasa Narayana acknowledge the financial support from Institution of Eminence (IoE-BHU) for Seed grant and Trans disciplinary research grant (TDR).

2

Amplification of the Mitochondrial Cytochrome Oxidase-1 (COX 1) Gene by the Polymerase Chain Reaction (PCR)

Varun Arya[1], Twinkle Sinha[1] and Saniya Tyagi[2]

[1]Department of Entomology & Agricultural Zoology, Institute of Agricultural Sciences, Banaras Hindu University, Varanasi - 221005, Uttar Pradesh, India

[2] BRD PG College, Deoria – 274001, Uttar Pradesh, India

Polymerase Chain Reaction (PCR) is a laboratory technique, developed by Kary Mullis in the 1980s, used in molecular biology to amplify and replicate a specific segment of the DNA (Mullis, 1990). This reaction in performed in a thermal cycler, involving three main steps: denaturation, annealing and extension.

Principle: PCR is based on the principle of selectively amplifying a specific region of DNA through a series of temperature-controlled cycles. The key components involved in PCR include the target DNA sequence, oligonucleotide primers, DNA polymerase enzyme (Taq DNA polymerase), deoxyribonucleotide triphosphate (dNTPs) (A, T, G, and C), Mg^{2+} and a thermal cycler apparatus. Initially, the double-stranded DNA is split into single strands under high temperature of 90-95 ^{0}C known as the denaturation step. It is followed by cooling the solution to a temperature that allows proper attachment of the oligonucleotide primers to its complementary sites in the strands, known as the annealing step. The DNA polymerase than synthesize new strands complementary to the template DNA with the help of dNTPs in the solution, known as the extension step. Mg^{2+} are essential cofactor for the DNA polymerase enzyme, essential for the synthesis of new strands. The newly synthesized DNA will then act as a template for further amplification.

Equipments required: Thermal cycler, EmeraldAmp® GT PCR master mix (Takara, Japan. Cat. No. RR310) (2X pre mix and nuclease-free water), universal barcode primers: LCO 1490 (5' GGTCAACAAATCATAAAGATATTGG 3') & HCO 2198 (5' TAAACTTCAGGGTGACCAAAAAATCA 3') (Folmer *et al.*, 1994), template DNA, PCR tubes (0.2 ml), PCR tubes mini-cooler/ice bucket, pipettes, micro tips, tip boxes, mini-centrifuge and gloves.

Protocol

- Perform the calculations for the different components as per the final volume of the PCR reaction, as per mentioned in Table 1.

  ***Note:** Always perform the calculations keeping a spare of 1-2 reaction to compensate the loss of chemicals during transfer (Human error in case large reactions).
- Put on a fresh pair of gloves, mask and clean lab coat (if available).
- Clean the working area, pipettes, tip boxes, tube bottles and tube stand with 70% ethyl alcohol at the beginning.
- Label the tubes clearly.
- Prepare the reaction master mix of the desired volume by mixing the PCR master mix, nuclease-free water along with both forward and reverse primers, as mentioned in Table 1.
- Centrifuge the solution by slight spinning the reactions at 10,000 rpm for 30-40 seconds.
- Transfer the calculated volume of the solution to their respectively labelled PCR tubes.
- Add required amount of the DNA template to the solution and centrifuge the tubes slightly.
- Switch-on the thermal cycler, open the lid and fit the reaction tubes in the sample groves.
- Adjust the thermal cyclic conditions as mentioned in Table 2.

Table 1. Amount of individual components for PCR solution for different volumes of the reaction mixtures.

Volume per reaction (in µl)	Volume of individual components (in µl)	
	PCR recipe in case individual component of PCR is used	
25	Nuclease-free water	16
	10X PCR buffer	2.5
	25 mM Mgcl2	2
	0.5 mM dNTPs,	0.5
	Taq DNA polymerase (5U/µL	0.1-0.2
	Forward and Reverse primers	0.5 µl each
	Template DNA	2
	PCR recipe in case PCR master mix is used	
25	PCR master mix	12.5
	Nuclease-free water	9.5
	Forward primer	0.5
	Reverse primer	0.5
	Template DNA	2
20	PCR master mix	10
	Nuclease-free water	7
	Forward primer	0.5
	Reverse primer	0.5
	Template DNA	2
10	PCR master mix	3.4
	Nuclease-free water	5
	Forward primer	0.3
	Reverse primer	0.3
	Template DNA	1

Table 2. Thermal cyclic conditions to amplify the COX 1 gene region (Srinivasa *et al.*, 2020).

Thermal cycling steps	Temperature (°C)	Duration
Initial denaturation (Hot start)	95	5 min
Denaturation	94	30 s }
Annealing	47	40 s } (35 cycles)
Extension (Initial)	72	40 s }
Extension (Final)	72	8 min
Cooling	4	Infinite (∞)

***Note:** The annealing temperature of the thermal cyclic condition for the COX 1 gene region is 47 ^{0}C, which will vary according to the primer pairs used. It is advised to read through the literature before assigning the thermal cyclic conditions for the reactions.

- Adjust the volume of the solution and press 'run'.
- After the reaction time is over, store the PCR products at -20ºC for further analysis.
- While mixing PCR mixture always add the larger amounts first them followed by the smaller amounts such as PCR master mix first followed by nuclease free water (NFW) or double distilled water etc.
- The recommended protocol is first PCR master mix followed by NFW and then primers. Mix it well by centrifuge then transfer the desired content to new labelled PCR tubes. Finally add the required quantity of template DNA.

References

Folmer O, Black M, Hoeh W, Lutz R, & Vrijenhoek R. (1994) DNA primers for amplification of mitochondrial cytochrome c oxidase subunit I from diverse metazoan invertebrates. *Molecular Marine Biology and Biotechnology*, *3*(5), 294-299.

Mullis, K. B. (1990). The unusual origin of the polymerase chain reaction. *Scientific American*, *262*(4), 56-65.

Srinivasa, N., Chander, S., & Chandel, R. K. (2020). Genetic homogeneity in brown planthopper, *Nilaparvata lugens* (Stål) as revealed from mitochondrial cytochrome oxidase I. *Current Science*, 6, 00113891.

Tyagi, S., Narayana, S., Singh, R. N., & Gowda, G. B. (2022). Molecular insights into wing polymorphism and migration patterns of rice planthoppers. In *Genetic Methods and Tools for Managing Crop Pests* (pp. 449-460). Springer, Singapore.

3

Separation and Visualization of DNA by Agarose Gel Electrophoresis

***Saniya Tyagi*[1], *Varun Arya*[2] *and Vishrutha C*[2]**

[1]*BRD PG College, Deoria – 274001, Uttar Pradesh, India*

[2]*Department of Entomology & Agricultural Zoology, Institute of Agricultural Sciences, Banaras Hindu University, Varanasi - 221005, Uttar Pradesh, India*

Agarose gel electrophoresis is used to resolve DNA fragments on the basis of their molecular weight. it will be visulaized by exposing to UV light.

Principle: As the phosphate backbone of DNA is negatively charged, the basic principle is the migration of DNA molecules from negative (cathode) to positive (anode) charge under constant electric field. The DNA will be separated mainly on the basis of their size, DNA conformation, concentration of gel and voltage. The distance travelled by the DNA molecules tell us about their molecular weight, where smaller DNA fragments will travel farther than the larger ones due to constant mass/charge ratio of DNA. The separated DNA molecules will be then visualised under UV light in gel documentation system after staining with suitable dye.

Equipments required: 50X Tris-acetate-EDTA (TAE)/Tris-borate-EDTA (TBE) buffer, distilled water, agarose powder, ethidium bromide (EtBr), gel loading dye, molecular ladder, weighing balance, spatula, magnetic stirrer, microwave oven, sterilized conical flask, measuring cylinder, blue-cap glass bottle, heat-resistant gloves, pipettes, micro tips, tip boxes, gel casting tray, gel comb, buffer tank, electrophoresis power supply, gloves and mask.

Protocol

- Put on a fresh pair of gloves, mask and clean lab coat (if available).
- Clean the working area, pipettes, tip boxes, tube bottles and tube stand with 70% ethyl alcohol at the beginning.
- Working solution of the EtBr must be prepared before starting the experiment. Weigh 1 g of the dye in a weighing balance and add it to 100 ml of sterilized distilled water.

- Stir the solution on a magnetic stirrer for 1-2 hours for proper dissolution of the dye. Wrap the container with an aluminium foil. This will be the EtBr stock solution.
- For making 2 ml of the working solution of 0.5 μg/ml EtBr, add 4 μl of the stock solution to 2 ml of distilled water in a tube.
- Vortex the tube for a few seconds, cover it with aluminium foil and store in the dark. This will be the working solution of EtBr.
- Prepare 1X TAE/TBE buffer solution by mixing 1 parts of 50X TAE/ TBE buffer to 49 parts of sterilized distilled water.

  ***Note:** For making 1 litre of 1X TAE/TBE buffer solution, we need to add 20 ml of 50X TAE/TBE buffer into 980 ml of sterilized distilled water. The solution should be stored in a blue-cap glass bottle and labelled properly.
- Weigh the desired amount of agarose powder in an electric balance and transfer it in a conical flask with the help of a clean spatula. Usually 0.5%-2% w/v gel solution is required.

  ***Note:** The concentration of agarose gel will depend upon the amplicon size of the sample that has to be visualized. Larger the size of amplicon, lower will be the concentration of gel (usually ranging between 0.5-2%).
- Add running TAE/TBE buffer to the flask. Do not swirl it as this will lead to agarose powder stick to the walls of the flask not in the solution.
- Heat up the mixture in microwave oven to melt the agarose till a clear solution is obtained. This is generally achieved in 90 to 150 seconds.

  ***Note:** Heat the flask at intervals of 30 seconds. This will lead to proper heating and will prevent the chances of spillage through boiling. Always keep an eye on the flask while heating.
- Take the flask out of the microwave by wearing proper heat-resistant gloves and allow the gel solution to cool down for 2-5 minutes at room temperature. After that add staining dye, usually ethidium bromide is used at a concentration of 0.5 μg/ml. Swirl the solution gently for proper mixing of the dye.

  ***Note:** Ethidium bromide is a suspected carcinogen. Hence, proper precautions must be taken while handling and disposing this chemical. It must be added as per the volume of gel solution (1.5-3 μl for 100 ml of gel).

  Note: You can replace ethidium bromide other with suitable dye such as SYBR™ Safe DNA Gel Stain which is less hazardous alternative.

Pour this mixture slowly in a gel casting tray and insert comb. The comb will help in making wells for the loading of the DNA.

- Allow the gel to set by keeping it at room temperature for 20-30 minutes.
- Once the gel is set slowly remove comb and keep gel along with casting tray in buffer solution.
- Add 1X TAE/TBE buffer solution enough to immerse the gel. Same buffer should be used when used for making gel.
- Add the loading dye/tracking dye (bromophenol blue) to the DNA. This dye will help to track the distance covered by DNA. Alternatively, PCR master mix embedded with loading dye are also available in market.
- Carefully load 2-5 µl DNA in the wells developed after removing comb from gel.
- Don't forget to add ladder. This will help in quantifying the DNA size.
- Close the lid and plug electrodes into the correct slots. Black electrode represents cathode and red electrode represents anode.
- Set up electrophoresis unit at desired voltage.

  ***Note:** The voltage and time will depend upon the concentration and the length of the gel. Extremely high voltage can sometimes melt the gel because of the heating effect and extremely low voltage will slow down the movement of DNA fragments. It is recommended to set the voltage at 70-80V for 60 minutes (0.5-1% gel)/120 minutes (1.5-2% gel) for a complete gel. Reduce the time to half for two rows of comb in a single gel.

- Turn on the power supply and check if the whole system is working. Bubbles arising from the electrodes when switched on will indicate the proper working of the system.

  ***Note:** Be cautious to observe the proper orientation of the gel in the buffer tank. The portion of the gel with wells must be towards cathode (black electrode).

- Run the gel till the dye has covered desired distance.
- Turn the power off and open the lid.
- Slowly take out the gel tray and drain off excess buffer from the gel.
- Turn on the gel documentation system and clean the loading surface with distilled water.

- Take out gel from the tray and put it on the loading surface.
- DNA bands can be seen after exposing to UV light
- Take out the gel and dispose it off along with buffer solution as per institution regulations.

Visualization of DNA

The amplified DNA of varying length will be shown in the form of bright bands corresponding to their sizes (bp) (see fig.). This will be confirmed by comparing them with the DNA ladder and the results can be later saved as photographs.

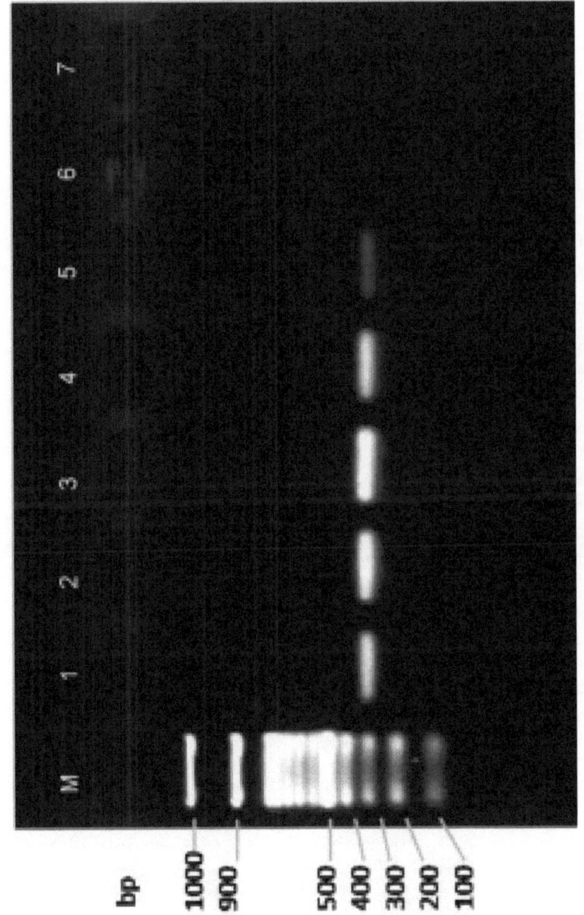

Fig. Gel image showing DNA bands at 307 bp from lanes 1 to 5. M represent 100 bp DNA ladder. Source: Tyagi *et al.* (2022).

Precautions

- Be very careful while handling ethidium bromide and dispose it off accordingly.
- Gel solution should be mixed carefully so that there will be no bubble formation.
- There should be a constant power supply during electrophoresis.
- Always wear gloves while handling ethidium bromide and keep it in a dark bottle to prevent any light inside it.
- Trays and combs should be cleaned with 90% ethanol and stored after drying.
- Keep changing buffer solution regularly.
- Avoid formation of primer dimer which can interrupt the results.

4

Quantification of the Nucleic Acids Using the Nano-Drop Spectrophotometer and Submission of COI Sequences in NCBI Genbank

Vishrutha C, Varun Arya and Srinivasa Narayana

Department of Entomology & Agricultural Zoology, Institute of Agricultural Sciences, Banaras Hindu University, Varanasi - 221005, Uttar Pradesh, India

Quantification of DNA is an essential step after its extraction in conjunction with the confirmation of the quality. NanoDrop spectrophometer is a powerful instrument in determining both purity and concentration of nucleic acids on the basis of absorbance values obtained by adding small amount of sample (1-2 μl) on the measurement surface.

Principle: The instrument works on the basic principle of spectrophotometry, which takes highly sensitive fluorescent measurements of the microvolumes. It takes use of microvolume sample retention technology based on surface tension property of the liquid and fibre optic technology. Together, these features make it possible to detect sample even in microlitres, which is much more than the traditional method that need cuvette-sized sample volume. The concentration of nucleic acid sample is normally determined by measuring its absorbance at 260 nm. In case of DNA, 260/280, 260/230, and 260/325 absorbance ratios are used to estimate DNA quality. These values indicate the presence of impurities in the DNA samples during the DNA extraction process.

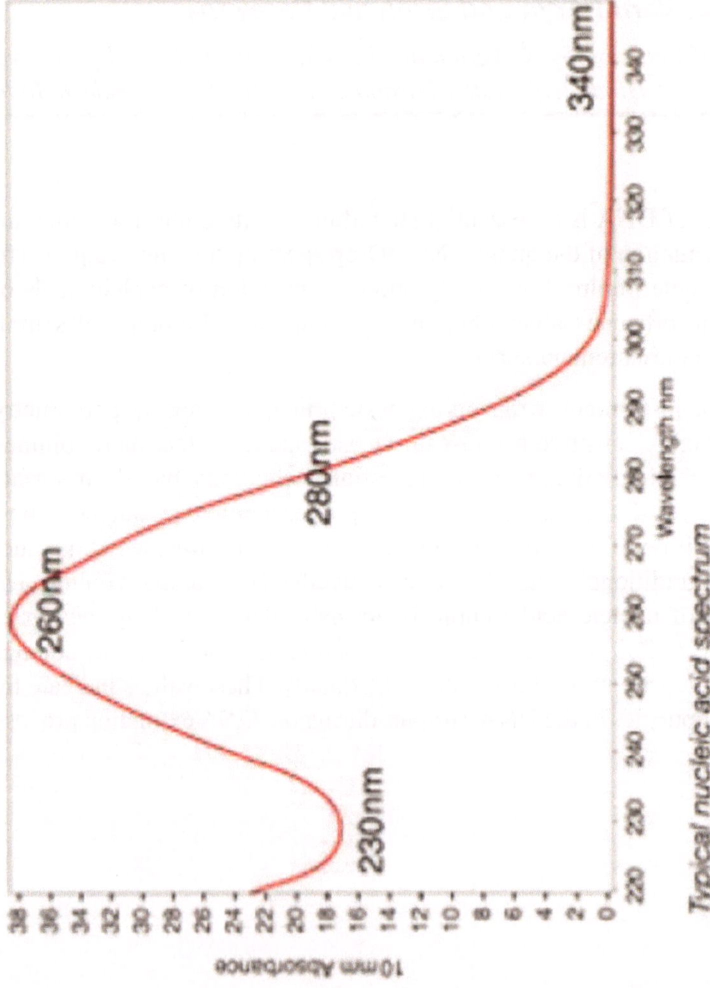

Fig. 1. A typical profile of a nucleic acid sample. Source: Desjardins & Conklin (2010).

Equipments required: NanoDrop spectrophotometer attached with a system, DNA sample kept in PCR mini-cooler/ice box, deionized water/elution buffer, pipettes, micro tips, clean tissue paper.

Protocol

- Take DNA samples as well as the soloution used for elution of DNA while extraction (usually elution buffer or nuclease-free water is used).
- Plug in and turn on the computer system attached with the NanoDrop spectrophotometer.
- Clean the lower and upper optical surfaces with 2-3 µl of clean deionized water and after that wipe it with lint free tissue paper.
- Take the blank by pipetting 1-2 µl of eluent/deionized water and then select "blank" on the screen.
- Gently lift and wipe the optical arm and pipette out 1-2 µl of DNA sample kept in ice box/PCR cooler on the arm pedestal.
- Gently lower the upper arm so that the sample get in touch with both the arms and click "measure" on the screen.
- Record concentration of DNA, spectral graph, as well as 260/280 and 260/230 absorbance values, which are the determinants of quality and purity of nucleic acids.
- Pure DNA sample would show 260/280 values near 1.8 and 260/230 values in the range of 2.0 to 2.2. Lower purity ratios may indicate the presence of phenol, salts, protein, or other such contaminants.
- While in RNA pure sample would show the 260/280 values near 2.0 and 260/230 values in the range of 2.0 to 2.2.
- Wipe the surface and shut down the system.

Submission of the DNA barcode sequences in the NCBI Genbank

The National Centre for Biotechnology Information (NCBI) was created in 1988, within the 'National Library of Medicine' at the 'National Institute of Health' for the development of the information system for molecular biology (Sayers *et al.*, 2024). NCBI GenBank is one of the 35 databases, all together contributing to 3.8 billion records of nucleotide sequences having different coding purposes. It is a database that contains nucleotide sequences of more than 30,000 organisms, classified up to the genus and species level (Sayers *et al.*, 2023). GenBank database has established standards and protocols for data submission, ensuring that sequences are accurately annotated and properly formatted. This helps maintain data integrity and ensures that the data can be effectively utilized by others. Submission of nucleotide sequences is an important step, as the sequences are examined, processed, and validated, before acceptance. Once accepted, it automatically generates a GenBank accession number, which is available to us and can be used further for various purposes. By depositing sequences in databases, one can make their data accessible to the broader scientific community. This fosters collaboration, facilitates reproducibility, and allows other scientists to build upon existing research. Databases serve as repositories for genetic and biological data. By submitting sequences, researchers ensure that their data is preserved and can be accessed in the future, even if the original research is no longer actively pursued. Many scientific journals require researchers to deposit sequence data in public databases as a condition for publication. This promotes transparency and ensures that the data supporting published research is accessible to others.

Steps involved in sequence submission in the NCBI GenBank database

- All the aligned sequences should be saved in fasta (>) format in text file before heading for submission.
- Open the 'NCBI Submission Portal' (https://submit.ncbi.nlm.nih.gov/) in a web browser.

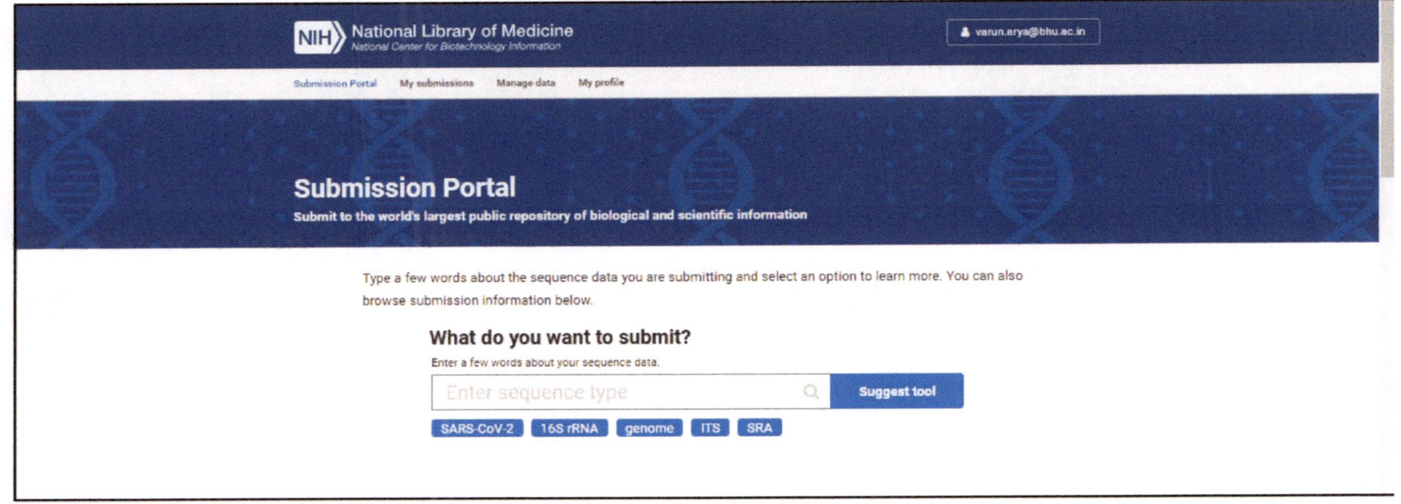

Fig. 2. NCBI sequence submission home page.

- Scroll down and select for submit under text box displaying 'Metazoan COX1'. Here, the selection may depend upon the type of nucleotide sequences submitted.
- Login with the 'Google Account'.
- Click on the 'New Submission' option.

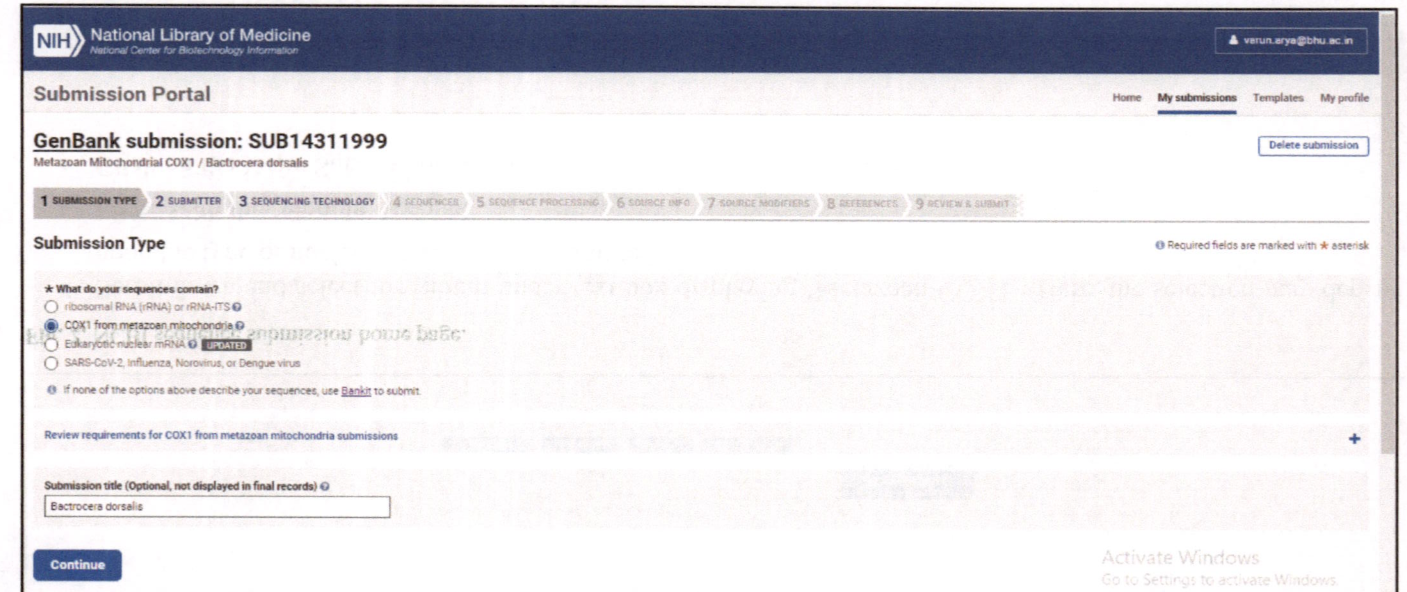

Fig. 3. Page representing NCBI sequence submission 'Submission type'.

- Fill in the details, such as 'Submission type: COX1 from metazoan mitochondria', 'Submission title', 'Submitter's details', 'Sequencing technology: Sanger dideoxy sequencing', 'Assembly state', 'Sequence upload', 'Sequence processing', 'Source info', 'References' and 'Review and submit'.

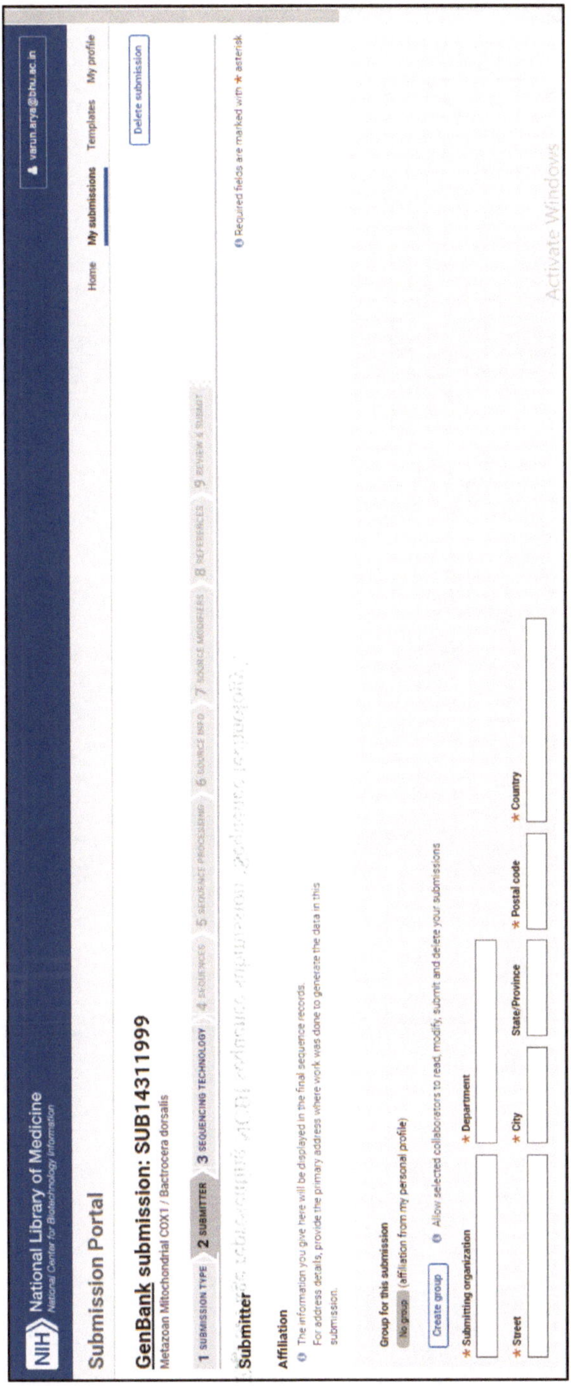

Fig. 4. Page representing NCBI sequence submission 'Submitter details'.

Fig. 5. Page representing NCBI sequence submission 'Sequence technology'.

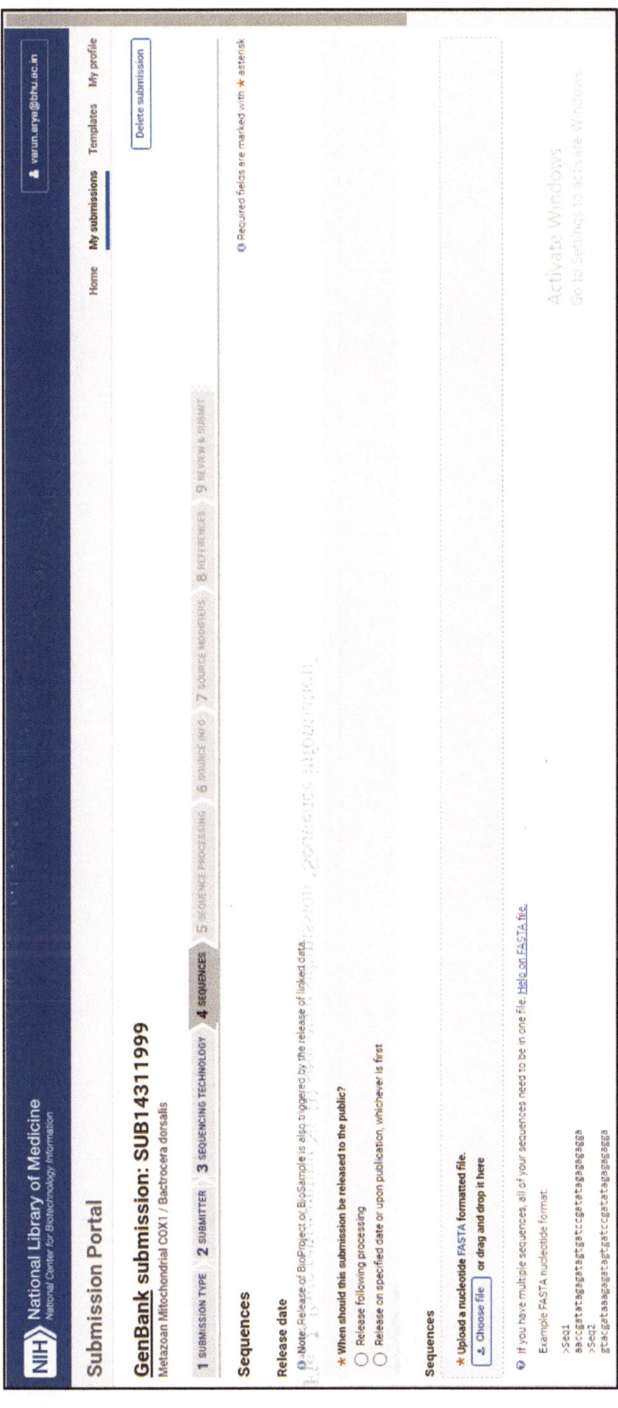

Fig. 6. Page representing NCBI sequence submission uploading sequences.

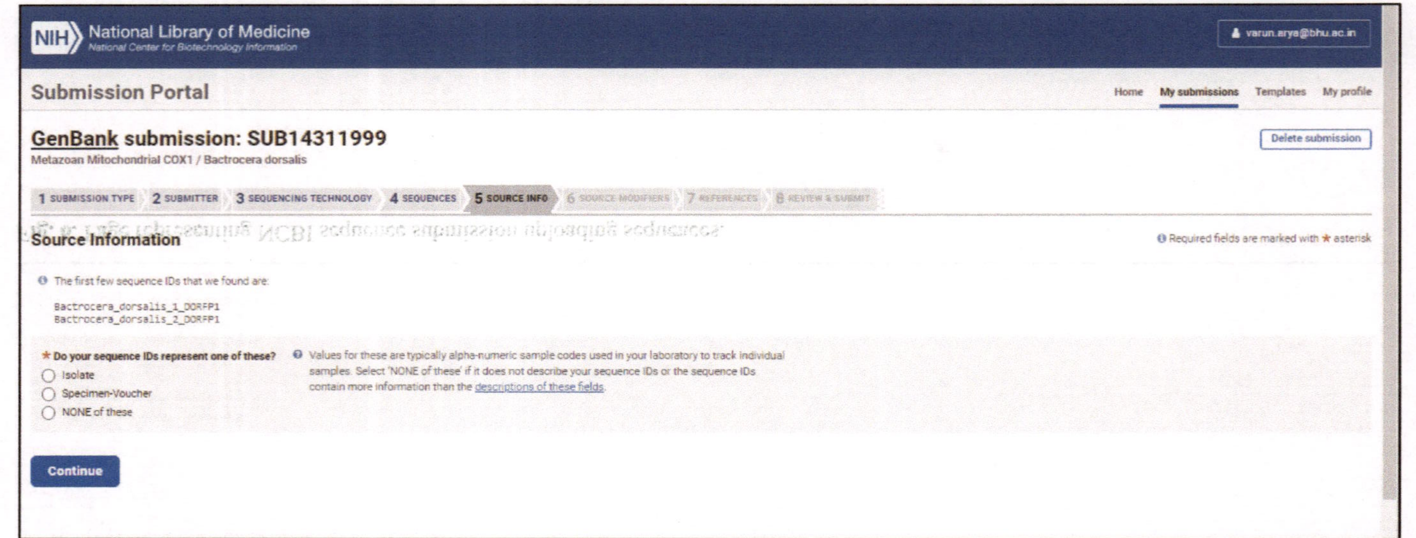

Fig. 7. Page representing NCBI sequence submission 'Sequence information'.

- Two or more than two sequences can be uploaded at once by making a single file containing all the sequences in fasta format. This file will be uploaded at the sequence submission step. The details of individual specimen should be given at the 'Source modifier' step.

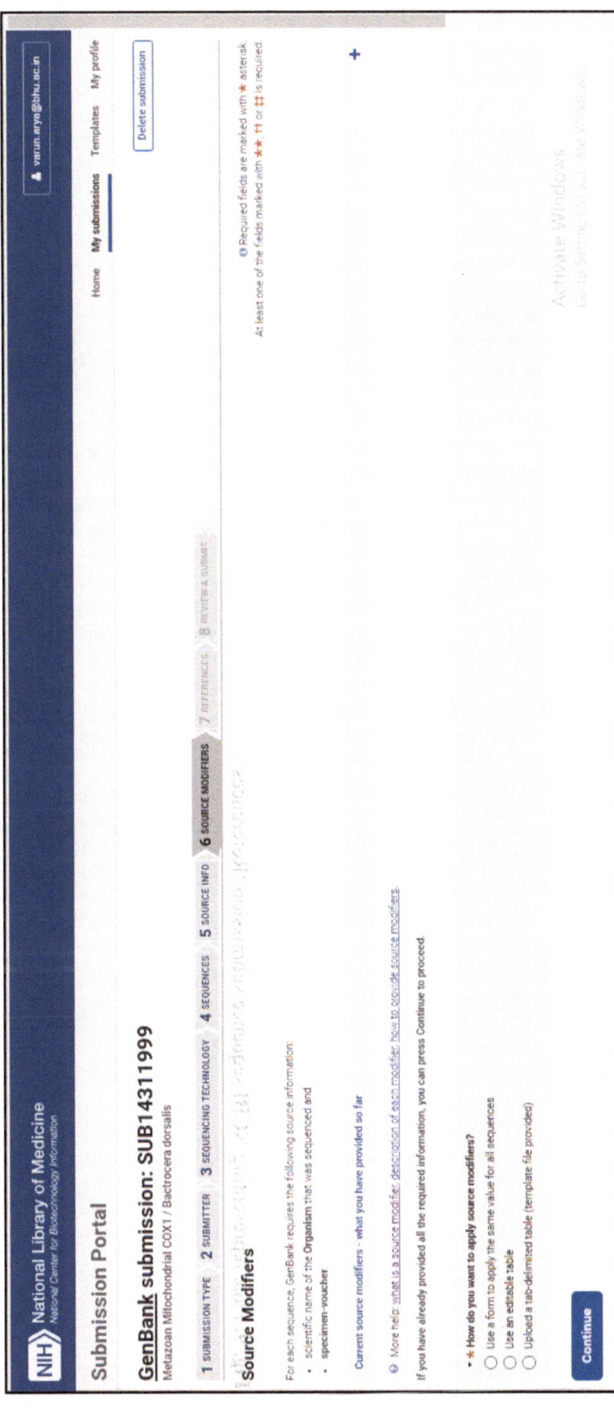

Fig. 8. Page representing NCBI sequence submission 'Source modifier'.

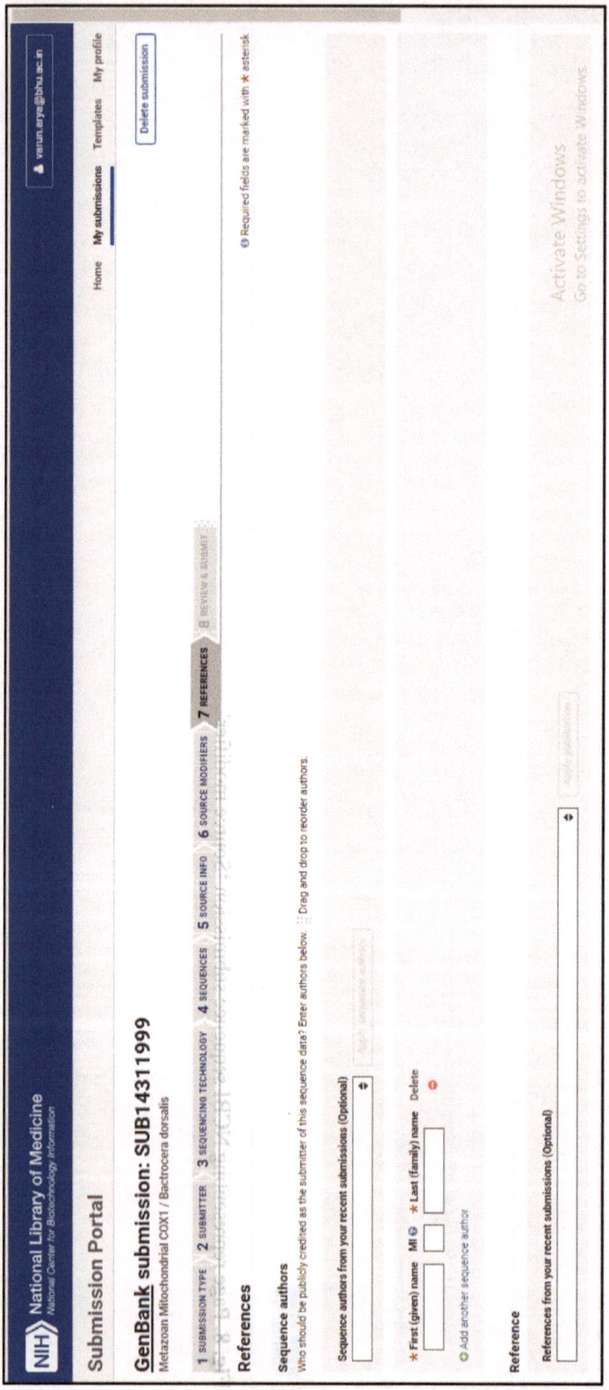

Fig. 9. Page representing NCBI sequence submission 'References'.

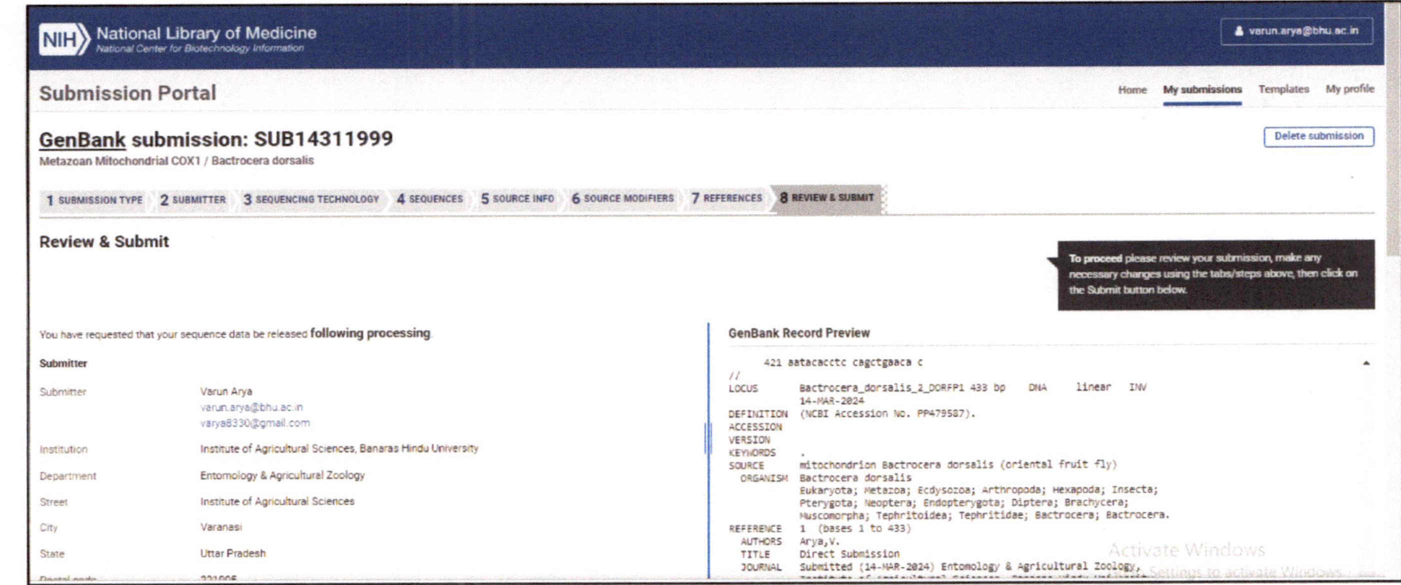

Fig. 10. Page representing NCBI sequence submission 'Review and submit'.

- Recheck the entries at the review step and click on 'Submit'.
- Any queries, details or the final accession numbers will be sent to the given email address, after processing.

References

Desjardins, P., & Conklin, D. (2010). NanoDrop microvolume quantitation of nucleic acids. *Journal of Visualized Experiments,* (45), e2565.

Sayers, E. W., Beck, J., Bolton, E. E., Brister, J. R., Chan, J., Comeau, D. C., & Sherry, S. T. (2024). Database resources of the national center for biotechnology information. *Nucleic Acids Research*, *52*(D1), D33.

Sayers, E. W., Bolton, E. E., Brister, J. R., Canese, K., Chan, J., Comeau, D. C., & Sherry, S. T. (2023). Database resources of the National Center for Biotechnology Information in 2023. *Nucleic acids research*, *51*(D1), D29.

5

DNA Extraction, Colony PCR and Evaluation of Biochemical Properties of Culturable Gut Bacteria from Insects

Vinay N[1] and Srinivasa Narayana[2]

[1]Faculty of Agricultural Sciences, GLA University, Mathura, Uttar Pradesh, India-281406

[2] BRD PG College, Deoria - 274001, Uttar Pradesh, India

[2]Department of Entomology & Agricultural Zoology, Institute of Agricultural Sciences, Banaras Hindu University, Varanasi - 221005, Uttar Pradesh, India

Aim: To identify culturable gut bacteria from insects.

Materials required: Micro scissors, Micro pestle, 1.5 ml centrifuge tubes, Pipettes, Petri plates, Inoculation loop, 0.5 ml PCR tube, Glass slides, non-absorbent cotton etc.

Instruments required: Autoclave, Refrigerator, Laminar Air Flow, Homogenizer, Biological Oxygen Demand (BOD) incubator, Thermal cycler, Gel Doc unit, Deep freezer pH meter, Spirit lamp.

Chemicals required: 70% ethanol, Phosphate Buffer Saline (PBS solution, pH 7.4), Distilled water, 0.01 M TE buffer, Forward primer - 27F (5'-AGAGTTTGATCCTGGCTCAG-3') and reverse primer - 1492R(5'-AAGGAGGTGATCCAGCCGCA), PCR Master mix, Nuclease free water, Agarose, Ethidium bromide, crystal violet, Grams's iodine, 95% ethanol, Safranin, 3% hydrogen peroxide, Tryptophan broth, Kovac's reagent, King's Base agar.

Procedure

1. Preparation of Nutrient Agar plates

- Weigh the required quantity of Nutrient agar and dissolve in clean conical flask containing required quantity of distilled water (13.00 g in 1000 ml of distilled water). Add agar at the rate of 2 g per 100 ml of distilled water and adjust the pH to 7.

- Cover the mouth of conical flask with cotton plug, wrap with newspaper and keep for autoclaving.
- After the autoclave, leave the media to reach room temperature and pour the media to petri plates under laminar air flow
- Invert the plates after one hour of pouring.

2. Isolation of gut bacteria

- Collect five numbers of actively growing, healthy larvae of the penultimate instar from their natural host crop (Any insect stage).
- Pre starve the larvae for 24 hrs to limit the risk of contamination caused by transient microbes
- Immobilize the larvae at 4ºC for 5 min for easy handling
- Disinfect the larval surface with 70% ethanol for 5 min to remove the surface contaminants
- Pin the larvae on the wax plates using sterilized insect pins
- Dissect the larvae under laminar air flow/under microscope quickly using sterilized micro scissors and separate the foregut, midgut and hindgut regions.
- Transfer the each gut region or whole gut region to a 1.5 ml centrifuge tube. Then add the 1.0 ml of Phosphate Buffer Saline (PBS solution, pH 7.4) and homogenise it.
- Make the serial dilutions for all the three guts up to 10^{-6} and from each dilution take 100 µl of aliquot and spread on Nutrient Agar plates
- Repeat the same procedure for remaining two replications.
- Incubate all the plates at 28 ± 2°C for 48 hrs in Biological Oxygen Demand (BOD) incubator.
- Pick up the distinct colonies from the mother plate and streak on the fresh nutrient agar plate for sub culturing and assign them with convenient code.
- Transfer the sub cultured bacteria into slants containing nutrient agar and preserve at -20°C in deep freezer for further use.

3. Preparation of slants

- Dissolve required quantity of nutrient agar in conical flask containing distilled water (adjust the pH to 7). Mix thoroughly and pour on an average of 8 -10 ml to each test tubes.

- Cover the test tubes with cotton plugs and keep for autoclaving.
- After autoclaving, keep the test tubes at 45° angle for overnight without disturbance.
- Because the 45° angle provide surface for growth of bacteria.

4. DNA Extraction

- Transfer a loopful of 24 hrs fresh culture to 0.5 ml PCR tube containing 100 µl of 0.01 M TE buffer.
- Incubate the PCR tubes in thermal cycler at 95ºC for 10 min followed by freezing at 4ºC for bursting of cells and to release DNA into TE buffer.

5. Colony PCR and 16S rDNA sequencing

- Use bacterial primers such as 27F (5'-AGAGTTTGATCCTGGCT-CAG-3') and 1492R (5′-AAGGAGGTGATCCAGCCGCA) for PCR amplification.
- Prepare PCR reaction mixture for 25 µl consist of 10 µl PCR master mix, 12.4 µl of nuclease free water, 0.3 µl of each primer and 2 µl of template DNA.
- Maintain PCR conditions as given below

Cycle Step	Temperature (ºC)	Time	Cycles
Initial denaturation	95	4 min	1
Denaturation	95	30 sec	36
Annealing	56	30 sec	
Extension	72	2 min	
Final extension	72	10 min	1

- Load the PCR products of 3 µl on 1.5% (w/v) agarose gel containing ethidium bromide (4µl of 1µg/ml) at 50 mV and visualized in GelDoc.
- Check the presence of distinct band from the samples and send the PCR products for sequencing.
- Check the quality of sequences and trim with Bio Edit software and submit the sequences to NCBI Gen bank to obtain accession number and to identify the bacterial species.
- Alternatively you can also identify the species at https://www.ezbiocloud.net/

6. Biochemical properties

a. Gram staining

- Smear the 24 hrs fresh bacterial culture on clean glass slide using sterile inoculation loop and allow to air dry
- Heats fix the bacterial smears by passing through gentle flame
- Add few drops of crystal violet on bacterial smears for 30 seconds and wash with distilled water
- Add few drops of Grams's iodine solution on bacterial smears for 60 seconds and wash with 95% ethanol by pouring it drop by drop until no more colour flows from the smear.
- Wash the smear with distilled water.
- Add few drops of safranin on bacterial smears for 30 seconds and wash with distilled water
- Air dry the slides and observe under microscope.
- The gram positive bacteria will retain the purple colour and gram negative strains do not retain the dye and appear pink colour.

b. Catalase test

- Select the colonies of 18 to 24 hrs fresh to test organisms on a sterilized glass slide using inoculation loop.
- And few drops of 3% hydrogen peroxide on it.
- Observe for bubbling.
- Consider the bacteria as catalase positive if bubbles appeared or consider negative if the bubble do not appear.

c. Indole test

- Select the colonies of 18 to 24 hrs and inoculate on 4 ml of tryptophan broth in a test tube.
- Incubate the test tubes at 28 ± 2°C for 24-28 hrs in shaker incubator.
- Add 0.5 ml of Kovac's reagent to each test tube.
- Observe for ring formation at the top layer.
- Consider the bacteria as indole positive if ring appears or consider as negative if ring is not formed.

d. Fluorescence test

- Streak the bacterial cultures of 24 to 48 hrs on King's Base agar media and keep the plates at 28 ± 2 °C for 48 hrs in Biological Oxygen Demand (BOD) incubator.
- Expose the cultures to UV light and observe for glowness of the culture.
- Consider the bacteria as fluorescence positive if the culture glows on media plate or consider as negative if bacteria does not glow.

e. Cellulolytic activity test

- Streak the colonies of 18 to 24 hrs old on corboxymehyl cellulose plates containing KH_2PO_4 – 0.5 g, $MgSO_4$- 0.25 g, CMC Na salt – 2 g, Agar – 15 g, Congo red – 0.4 g, Gelatin – 2 g per liter of distilled water (adjust pH to 7).
- Incubate the plates at 28 ± 2°C in Biological Oxygen Demand (BOD) incubator for 10 days.
- Consider the bacteria as cellulolytic if halo zone appears or consider as non-cellulolytic if halo zone doesn't appears.

Acknowledgement

Dr. Srinivasa Narayana acknowledge the financial support from Institution of Eminence (IoE-BHU) for Seed grant and Trans disciplinary research grant (TDR).

6

Immunoblotting

Sriparna Pal and Rakesh Verma

Department of Zoology, Institute of Science, Banaras Hindu University, Varanasi - 221005, Uttar Pradesh, India

Immunoblotting or Western blotting is a technique employed to observe proteins separated through gel electrophoresis. After the proteins migrate from the gel to a nitrocellulose or PVDF membrane due to an electrical current, the membrane is exposed to antibodies tailored for the desired target. Subsequently, secondary antibodies and detection reagents are utilized to visualize the proteins of interest on the membrane.

1. Preparation of tissue homogenate

The initial phase of sample preparation involves extracting proteins from their origin. Typically, proteins are extracted from cells or tissues through a process called lysis. This procedure disrupts the cell membrane to separate proteins from the insoluble components of the cell. Different lysis buffers are available for sample preparation in Western blotting and they differ in the potency of their detergents used to liberate soluble proteins.

Lysate buffer

It is used for whole cell extract, both nuclear and membrane-bound protein. It contains ionic detergents that solubilize the protein easily.

Components of lysate buffer

I. RIPA (Radio immuno precipitation assay) → Storage is 4°C.

   a) PBS – 49 mL

   b) NP40 – 500 μL (cationic detergent)

   c) 10% SDS – 500 μL (anionic detergent)

   - NP40 (Nonyl phenoxypolyethoxy ethanol) breaks down nuclear and cytoplasmic membrane. Storage is 4°C.

II. PMSF (Phenyl methyl sulphonyl fluoride) → Storage is -20ºC.

- Light sensitive. Should be prepared in dark. It prevents the action of cysteine and tyrosine proteases.
- 10 mg PMSF powder in 1 mL absolute alcohol (10 mg/mL)

III. NaOV (Sodium orthovanadate) → Storage is -20ºC.

- Inhibitor of protein tyrosine phosphatase, alkaline phosphatase and ATPase.

IV. Aprotinin → Storage is -20ºC.

- Inhibitor of serine protease and prevents action of tyrosine hydroxylase.
- Light sensitive. Should be prepared in dark.
- Prepared in milli Q water (10 mg/mL).

Components of lysate buffer

Lysate buffer	For 1 mL
RIPA	970 µL
NaOV	20 µL
PMSF	10 µL
Aprotinin	1 µL

Procedure

Dissect the tissue using sterile tools, preferably on ice, and work swiftly to prevent degradation by proteases

↓

Immediately after dissection, snap freeze the tissue by placing it in microcentrifuge or Eppendorf tubes and store the samples at -80°C for future use or keep them on ice if homogenization is immediate

↓

Prepare 10% tissue homogenate (For approximately a 100 mg tissue sample, add about 1000 µL of ice-cold lysis buffer to the tube)

↓

After homogenization, incubate the samples on ice for 1 hour

↓

After the incubation period, centrifuge the lysate for 30 minutes at 12000 rpm at 4°C in a microcentrifuge. Carefully remove the tubes from the centrifuge and place them on ice

↓

Aspirate the supernatant from the centrifuged lysate and transfer it to a fresh tube kept on ice, discarding the pellet. This supernatant contains the protein extract and can be further processed or stored appropriately

↓

Proceed for protein estimation and store the rest at -80°C.

2. Bradford assay for protein estimation

The Bradford assay is a commonly used method for estimating the concentration of proteins in a sample. The Bradford assay relies on the binding of the dye Coomassie Brilliant Blue G-250 (CBB-G) to proteins. At low pH, CBB-G dye is reddish brown in color. On interaction with the amino acid residues in protein, it undergoes complex formation of which gives a stable blue adduct. The intensity of the blue color is directly proportional to the concentration of proteins in the sample. The dye binds to basic and aromatic amino acid residues in proteins, leading to the color change.

***Principle*:** CBB-G (red) donates electrons to ionizable groups of proteins which leads to disruption of the native state and exposure of the hydrophobic residues. These hydrophobic groups on tertiary structure interact with Van der Waal's forces such that negatively charged residues are positioned close to positively charged ones leading to formation of a stable complex which gives A_{max} at 590 nm.

This method actually measures the presence of basic amino acid residues, i.e., arginine, lysine, histidine which contribute to formation of protein-dye complex.

For standard

Bovine serum albumin (BSA) → 2 mg/mL PBS (stock)

Make 20 times diluted lysate → 25 µL lysate + 475 µL PBS

	Standard	Concentration
S1	40 µL stock	2000 µg/mL
S2	32 µL S1 + 8 µL diluted lysate	1600 µg/mL
S3	20 µL S2 + 20 µL diluted lysate	800 µg/mL
S4	20 µL S3 + 20 µL diluted lysate	400 µg/mL
S5	20 µL S4 + 20 µL diluted lysate	200 µg/mL
S6	20 µL S5 + 20 µL diluted lysate	100 µg/mL
S7	20 µL S6 + 20 µL diluted lysate	50 µg/mL

For experimental samples

- Dilute the sample 20 times with PBS → 2 μL sample + 38 μL PBS

For blank

- Use diluted lysate

Procedure

Mark the 96 well plate according to blank, sample and standard

↓

Run in triplicates

↓

Load 5 μL of standard or sample or diluted lysate (for blank) in respective wells

↓

Add 200 μL of Bradford reagent in each well

↓

Blue color develops in each well except for blank

↓

Take OD at 590 nm

Calculation

Protein content = 1 OD (from standard) x Experimental OD x Dilution factor (20)

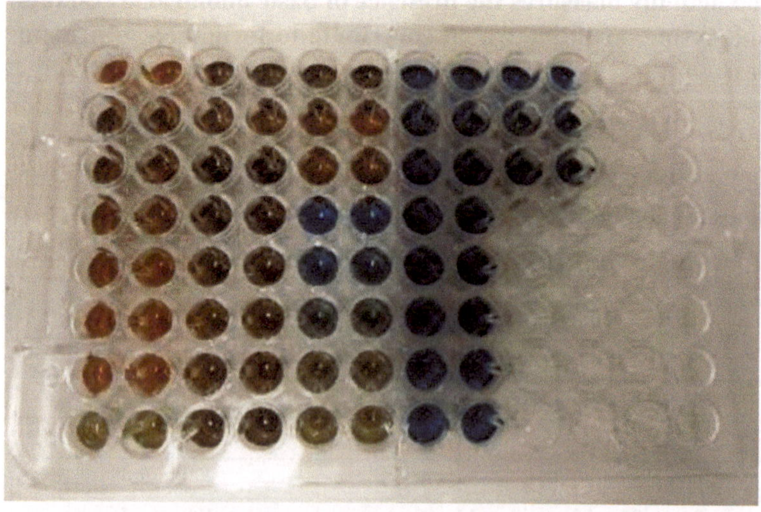

Fig 1: Protein estimation by Bradford method

3. Electrophoresis of Proteins

Principle: The first stage of Western blotting involves physically separating proteins within a gel matrix through a process known as gel electrophoresis. Sodium dodecyl sulfate—polyacrylamide gel electrophoresis (SDS–PAGE) is the most widely used method to separate proteins based on their molecular mass. In SDS-PAGE, proteins are denatured and coated with SDS, which imparts a uniform negative charge to each protein molecule. This treatment disrupts the native structure of proteins and confers a uniform charge-to-mass ratio, allowing proteins to migrate through the gel primarily based on size. Cross-linked polyacrylamide gels are created through the polymerization of acrylamide monomers with smaller quantities of N, N'-methylene-bisacrylamide. Bis-acrylamide serves as a cross-linking agent, essentially linking two acrylamide molecules *via* a methylene group. The polymerization of acrylamide is initiated through free-radical catalysis, triggered by the addition of ammonium persulfate (APS) acting as the initiator and the base N, N, N', N'-tetramethylenediamine (TEMED) acting as a catalyst.

Discontinuous gels, also known as Laemmli gels, are widely utilized in gel systems due to their ability to yield sharper and more defined bands compared to continuous gel systems. This setup comprises of two layers of gels stacked on top of each other, each with distinct acrylamide concentrations and pH levels. The disparity in pH between the two gel layers and the buffer influences the mobility of ions, particularly zwitterions, across the gels, consequently affecting the migration of proteins. The upper layer of the gel, referred to as the **stacking** or **focusing gel**, contains a fixed low percentage of acrylamide (generally 5%) and maintains the lower pH level (pH of 6.8). The primary function of the stacking gel is to concentrate all proteins into a singular, tightly packed band before they progress into the lower region of the gel. This concentration enhances the sharpness and clarity of individual protein bands in the resolving gel. The lower portion of the gel, termed the **resolving gel**, contains the necessary percentage of acrylamide required to separate the protein of interest and maintains the higher pH level (pH of 8.8). The resolving gel's purpose is to segregate proteins based on their size, and thus, the composition of the gel, including pH and acrylamide percentage, is tailored to ensure that proteins predominantly migrate through the gel according to their molecular weight. When an electrical current is applied, the proteins migrate through the gel matrix towards the positively charged electrode. Smaller proteins move faster and thus travel farther through the gel, while larger proteins migrate more slowly and remain closer to the origin.

Procedure

Cast the both gels. First the resolving gel and then the stacking gel.

↓

Take equal amounts (50–100 μg) of protein for each sample in microcentrifuge tubes and add 10 μL of sample loading buffer to each tube.

↓

Heat on dry bath at 100°C for 10 minutes and then immediately put in ice for 10 minutes.

↓

Load the samples in respective wells in stacking gel along with a molecular weight marker.

↓

Run the samples at 50 V till the samples enter the resolving gel. Then run at 100 V.

↓

After electrophoresis, the separated proteins can be visualized by staining the gel with dyes such as Coomassie Blue or silver stain.

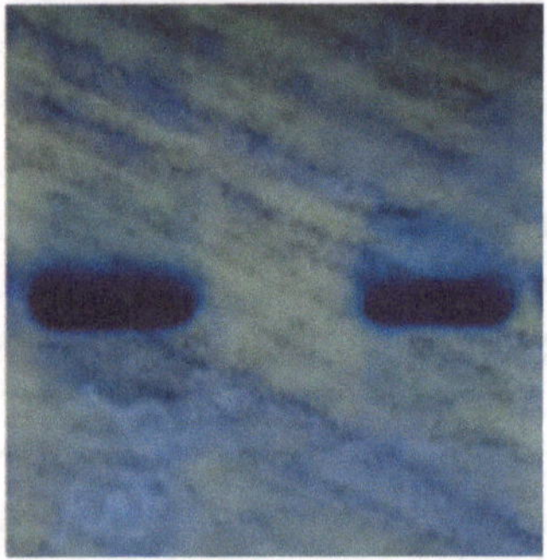

Fig 2: Visualization of protein bands in gel by Coomassie Brilliant Blue staining

Components to make gels and buffers:

Resolving gel:

	15%	12%	10%	8%
Distilled water	2.3 mL	3.3 mL	4 mL	4.6 mL
Acrylamide	5 mL	4 mL	3.3 mL	2.7 mL
Tris (pH 8.8, 1.5 M)	2.5 mL	2.5 mL	2.5 mL	2.5 mL
10% SDS	400 μL	400 μL	400 μL	400 μL
10% APS	400 μL	400 μL	400 μL	400 μL
TEMED	5 μL	5 μL	5 μL	5 μL

5% Stacking gel

Distilled water	3.4 mL
Acrylamide	830 μL
Tris (pH 6.8, 1 M)	630 μL
10% SDS	50 μL
10% APS	50 μL
TEMED	5 μL

Sample loading buffer

Tris (pH 6.8, 1 M)	100 μL
Glycerol	200 μL
10% SDS	400 μL
Distilled water	250 μL
β-marcaptoethanol	50 μL

Add a pinch of bromophenol blue to this above solution to turn it blue.

Running buffer (5X)

Tris base	9 g
Glycine	43.2 g
SDS	3 g
Distilled water	500 mL

Make 1X working running buffer from 5X (100 mL of 5X running buffer + 400 mL of distilled water).

4. Transfer of proteins from gel to membrane

Following the electrophoretic separation of proteins in the gel, the next step involves transferring these proteins onto a solid membrane support. Nitrocellulose (NC) and polyvinylidene difluoride (PVDF) are the two common membrane types used for Western blots, each with its own suitability depending on experimental setup and sample composition. Similar to gel electrophoresis, this transfer process utilizes electricity to guide negatively charged proteins towards the positively charged electrode. Transfer efficiency is influenced by factors such as the transfer apparatus, individual protein characteristics, transfer buffer composition and transfer conditions.

Fig 3: A typical semi-dry transfer setup

Procedure

Soak the stacks and membrane in transfer buffer for 10 minutes at 4°C before transfer. In case of PVDF, charge the membrane for 30 minutes in methanol.

↓

Transfer for 10-15 minutes (for semi-dry transfer method) or 12-16 hours (for wet transfer method).

↓

Take out the membrane and stain it with Ponceau S for 30 seconds to visualize the protein bands.

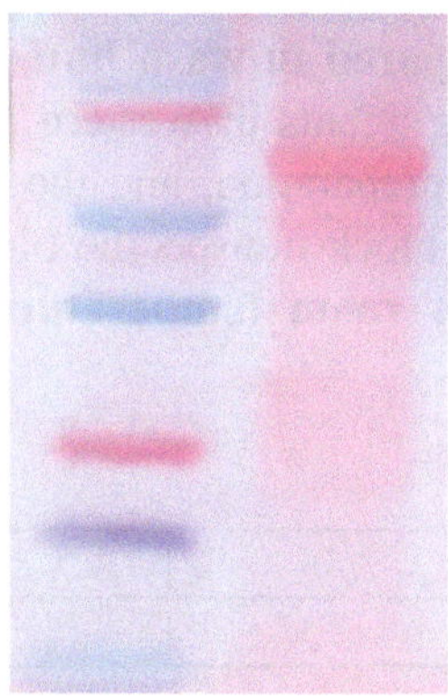

Fig 4: Visualization of protein bands in NC membrane by Ponceau S staining

Components to make buffers

Transfer buffer (semi-dry transfer method)

Transfer buffer (5X)	100 mL
Absolute ethanol	100 mL
Distilled water	300 mL

Transfer buffer (wet transfer method)

Glycine	7.13 g
Tris Base	1.514 g
Methanol	100 mL
Distilled water	400 mL

5. Membrane blocking

Although antibodies employed in Western blotting exhibit a strong affinity for particular proteins, they also possess a tendency to bind nonspecifically and with low affinity to the membrane used in the process. This nonspecific binding can lead to an elevated background signal during imaging, consequently diminishing detection sensitivity. To mitigate this issue and maintain a minimal background signal, the membrane is subjected to incubation in a blocking solution subsequent to transfer. The blocking solution functions by attaching to the nonspecific antibody binding sites on the membrane, thereby obstructing antibody binding through occlusion.

5% Non-fat dry milk prepared in wash buffer serves as one of the most commonly utilized blocking agents in Western blotting. It is favored for its cost-effectiveness, ease of preparation, and the presence of a diverse array of proteins that effectively block nonspecific binding. In most experiments, **blocking for one hour at room temperature** is typically adequate for achieving optimal results.

Wash buffer (PBST)

PBS	500 mL
Triton X-100	500 μL

6. Incubation of membrane and detection of antibody

Following the blocking step, the membrane is prepared for probing with antibodies.

Procedure

Incubate the membrane with the primary antibody specific to the target protein diluted in blocking buffer at 4°C for overnight.

↓

Next morning, bring the membrane to room temperature.

↓

Wash the membrane multiple times (usually 3 times, 10 minutes apart) with PBST (wash buffer) to remove unbound primary antibody.

↓

Incubate the membrane with appropriate HRP-conjugated secondary antibody diluted in blocking buffer.

↓

Incubate the membrane for 2 hours at room temperature with gentle agitation.

↓

Wash the membrane multiple times (usually 3 times, 10 minutes apart) with PBST (wash buffer) to remove unbound secondary antibody.

↓

Incubate the membrane with a chemiluminescent substrate according to the manufacturer's instructions.

↓

Visualize the protein bands using X-ray film or a digital imaging system.

↓

Analyze the resulting bands to determine the presence and intensity of the target protein.

Fig 5: Visualization of protein band in X-Ray film

X-ray developer

		Role
Metol	1.1 g	Reduces the silver in the X-ray film to metallic silver
Sodium sulphite	36 g	Prevents oxidation by aerial oxygen
Sodium carbonate	28 g	Maintains alkaline pH
Potassium bromide	2 g	Prevents chemical fogging
Hydroquinone	4.4 g	Recharges after each reaction

Hydroxyquinone is to be added at last when all other constituents dissolve properly. Add distilled water and make the volume to 500 mL.

Acknowledgement

Authors acknowledge the financial support from Council of Science and Technology, Uttar Pradesh (CSTUP), Lucknow; Institution of Eminence (IoE-BHU) for Incentive grant, Transdisciplinary research grant (TDR), Raja Jwala Prasad- PDF grant and Department of Health Research (DHR), New Delhi to Dr. Rakesh Verma.

References

Azure Biosystem. (n.d.). Western Blotting Guidebook. chrome-extension://efaidnbmnnnibpcajpcglclefindmkaj/https://www.bu.edu/picf/files/2019/05/Azure-Western-Blotting-Guidebook.pdf

Genei, A. (n.d.). Western Blot Guide Protocols.

Gwozdz, T., & Dorey, K. (2017). Western Blot. In Basic Science Methods for Clinical Researchers. Elsevier Inc. https://doi.org/10.1016/B978-0-12-803077-6.00006-0

Kumar, J., Haldar, C., & Verma, R. (2021). Melatonin Ameliorates LPS-Induced Testicular Nitro-oxidative Stress (iNOS/TNFα) and Inflammation (NF-kB/COX-2) via Modulation of SIRT-1. Reproductive Sciences, 28(12), 3417–3430. https://doi.org/10.1007/s43032-021-00597-0

Pal, S., Sahu, A., Verma, R., & Haldar, C. (2023). BPS-induced ovarian dysfunction: Protective actions of melatonin via modulation of SIRT-1/Nrf2/NFκB and IR/PI3K/pAkt/GLUT□4 expressions in adult golden hamster. Journal of Pineal Research, March, 1–15. https://doi.org/10.1111/jpi.12869

Sahu, A., & Verma, R. (2023). Bisphenol S dysregulates thyroid hormone homeostasis; Testicular survival, redox and metabolic status: Ameliorative actions of melatonin. Environmental Toxicology and Pharmacology, 104, 104300.

Wilson, K., & Walker, J. (2010). Principles and Techniques of Biochemistry and Molecular Biology (K. Wilson & J. Walker (eds.); Seventh). Cambridge University Press.

7

Real-Time or Quantitative Polymerase Chain Reaction

Sriparna Pal and Rakesh Verma

Department of Zoology, Institute of Science, Banaras Hindu University, Varanasi - 221005, Uttar Pradesh, India

The polymerase chain reaction (PCR) stands out as a highly potent technology within the realm of molecular biology. With PCR, particular sequences present within a DNA or cDNA template can undergo replication, termed "amplification," increasing their quantity manifold, ranging from thousands to millions. This process entails the utilization of sequence-specific oligonucleotides, a DNA polymerase capable of withstanding high temperatures, and thermal cycling.

Real Time PCR, a variant of the standard PCR technique, serves as a prevalent method for quantifying DNA or RNA within a sample. Through the use of sequence-specific primers, the quantity of copies of a specific DNA or RNA sequence can be assessed. This is achieved by monitoring the amount of amplified product at each stage throughout the PCR cycle. Earlier cycles exhibit amplification if a particular sequence (DNA or RNA) is abundant in the sample, while amplification in later cycles indicates scarcity of the sequence. Quantification of the amplified product is facilitated through fluorescent probes or DNA-binding dyes that emit fluorescence, coupled with real-time PCR instruments capable of measuring fluorescence while conducting the necessary thermal cycling for the PCR reaction.

Procedure

1. **Primer design:** To design and produce primers for the target gene, it is essential to ensure they are highly specific to the target sequences, capable of forming stable duplexes, and devoid of secondary structures. The primers should ideally be around 20-25 nucleotides long, with the GC content balanced as closely as feasible. The amplification length should fall within the range of 80-200 base pairs.

2. **RNA isolation:** RNA extraction involves isolating total RNA from cells or tissues of plants or animals using a conventional extraction protocol employing Trizol. The extracted RNA is then dissolved in DEPC-treated deionized water and quantified using a spectrophotometer. For RNAs possessing polyadenylated tails, enrichment is performed using an mRNA Purification Kit.

RNA isolation from tissue

Homogenize tissue (100 mg) in 1 mL TRIzol

↓

Let it stand for 5 min at RT

↓

Add 200 uL of chloroform / 1 mL TRIzol

↓

MIX THOUROUGHLY by inverting the tubes upside down

↓

Centrifuge at 12000 RPM for 15 min at 4ºC

↓

Transfer the aqueous phase into a new tube

↓

Add 500 uL of Isopropanol / 1 mL TRIzol

↓

Mix gently and incubate for 10 min at RT

↓

Centrifuge at 12000 RPM for 10 min at 4ºC

↓

Discard the supernatant

↓

Wash pelet with 1 mL of 75% ethanol / 1 mL TRIzol

↓

Centrifuge at 12000 RPM for 15 min at 4ºC

↓

Air-dry the pellet for 20 min at RT [time may vary a/c to pellet size]

↓

Dissolve the pellet in 20-50 uL DEPC water/Dnase or Rnase free water

↓

Keep at RT until completely dissolved

↓

Quantify the RNA and run gel

↓

Store at -80°C for further use

↓

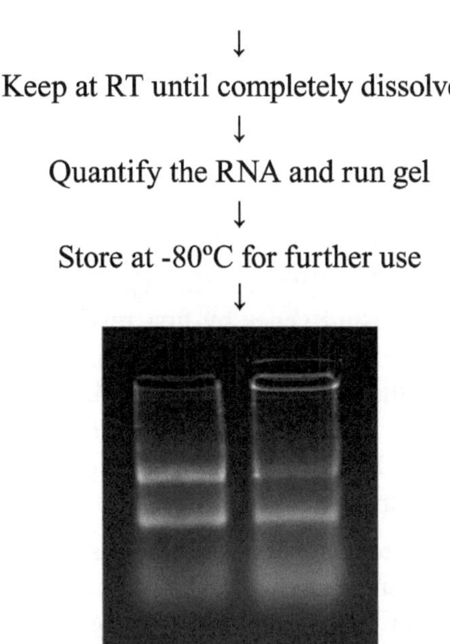

Fig 1: Visualization of RNA bands

For 1% denaturing agarose gel

	30 mL	40 mL	50 mL	80 mL	100 mL
Agarose	0.3 g	0.4 g	0.5 g	0.8 g	1 g
Formaldehyde	21. mL	2.8 mL	3.5 mL	5.6 mL	7 mL
MOPS/TAE Buffer (1XO)	3 mL	4 mL	5 mL	8 mL	10 mL
DEPC treated water	24.9 mL	33.2 mL	41.5 mL	66.4 mL	83 mL
EtBr	5 uL	5 uL	5 uL	5 uL	5 uL

Loading dye (2X for RNA)

Formamide	950 uL
10% SDS	2.5 uL
EDTA (0.5 M)	1 uL
EtBr (10 mg/ mL)	2.5 uL
Bromophenol blue	a pinch
Xylene cyanol	a pinch

TAE (Tris-Acetate-EDTA) Buffer 50X Stock Recipe (500 mL)

Transfer buffer (5X)	100 mL
Absolute ethanol	100 mL
Distilled water	300 mL

Dilute the TAE buffer to 1X before running

3. **Reverse transcription (RNA→cDNA):** RT-PCR is employed for the cloning of expressed genes by first reverse transcribing the RNA of interest into its DNA complement utilizing reverse transcriptase. Following this, the newly generated cDNA is amplified using conventional PCR. Total RNA serves as the template, with reverse transcription into cDNA achieved using Oligo (dT) or random primers along with reverse transcriptase.

4. **Real-time PCR:** Real-time PCR (qPCR) has emerged as the method of choice for quantifying gene expression, particularly for comprehensive array analysis and global gene expression studies. Presently, four distinct fluorescent DNA probes are utilized for detecting real-time PCR products: SYBR Green, TaqMan, Molecular Beacons, and Scorpions. All these probes facilitate the detection of PCR products by emitting a fluorescent signal. However, their mechanisms differ: SYBR Green dye emits fluorescence upon binding to double-stranded DNA in solution, while TaqMan probes, Molecular Beacons, and Scorpions rely on Förster Resonance Energy Transfer (FRET) coupling between the dye molecule and a quencher moiety within the oligonucleotide substrates to generate fluorescence.

5. **Result analysis:** During the initial cycles of the qPCR amplification reaction, the fluorescence signal undergoes minimal change and approaches a linear trend, forming the baseline. The fluorescence signal observed in the first 15 cycles represents the background fluorescence signal. The fluorescence threshold is set in the exponential phase of the PCR amplification, typically at a value ten times the standard deviation of the fluorescence signals observed between cycles 3 and 15. This threshold defines the fluorescence domain.

   The cycle threshold (CT) value signifies the number of cycles required for the fluorescence signal to surpass the set fluorescence domain value. Lower CT values correspond to higher initial template copy numbers, indicating greater target abundance, and vice versa.

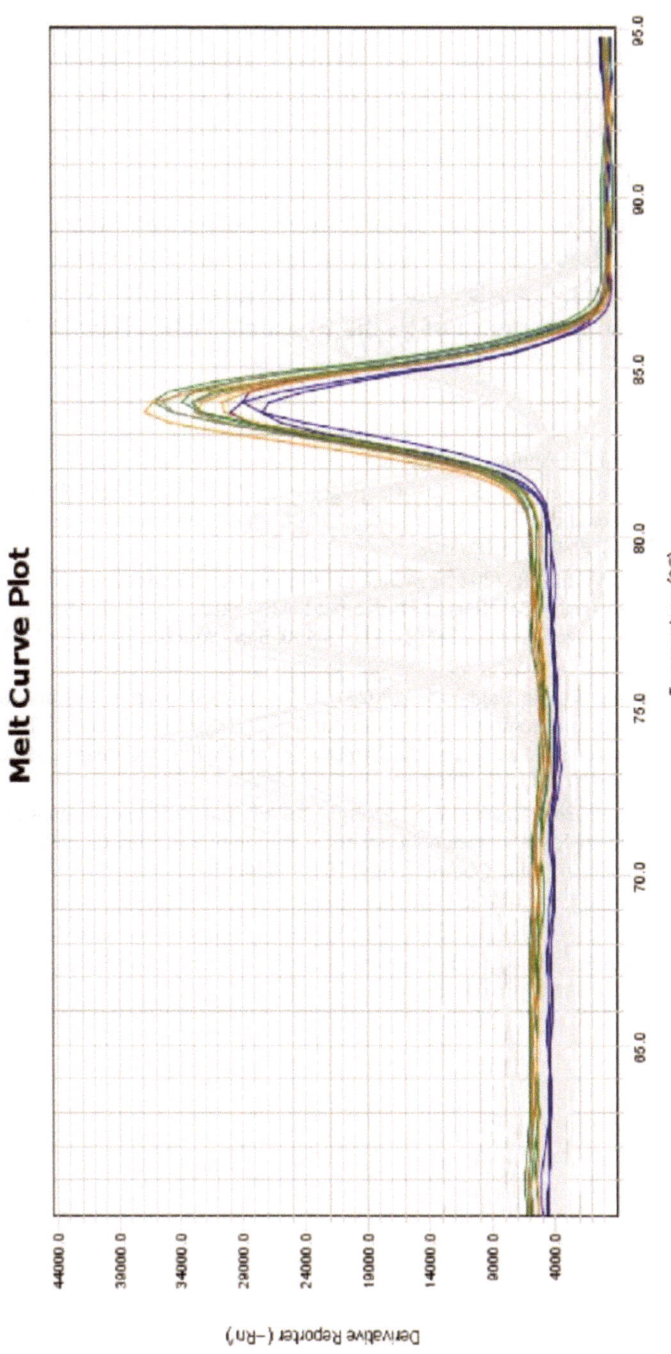

Fig 2: Graph representing melt curve plot

Achieving consistent amplification across a broad dynamic range is crucial in qPCR methodology. The $\Delta\Delta$Ct method is commonly employed for gene expression analysis, which involves comparing the difference in CT values between target and reference genes to determine relative expression levels.

Acknowledgement

Authors acknowledge the financial support from Council of Science and Technology, Uttar Pradesh (CSTUP), Lucknow; Institution of Eminence (IoE-BHU) for Incentive grant, Transdisciplinary research grant (TDR), Raja Jwala Prasad- PDF grant and Department of Health Research (DHR), New Delhi to Dr. Rakesh Verma.

References

BioRad. (n.d.). Real-Time PCR Applications Guide.

Fraga, D., Meulia, T., & Fenster, S. (2014). Real-Time PCR. In Current Protocols in Essential Laboratory Techniques (Vol. 2014, Issue February). https://doi.org/10.1002/9780470089941.et1003s08

Kumar, J., Haldar, C., & Verma, R. (2021). Melatonin Ameliorates LPS-Induced Testicular Nitro-oxidative Stress (iNOS/TNFα) and Inflammation (NF-kB/COX-2) via Modulation of SIRT-1. Reproductive Sciences, 28(12), 3417–3430. https://doi.org/10.1007/s43032-021-00597-0

Pal, S., Sahu, A., Verma, R., & Haldar, C. (2023). BPS□induced ovarian dysfunction: Protective actions of melatonin via modulation of SIRT□1/Nrf2/NFκB and IR/PI3K/pAkt/GLUT□4 expressions in adult golden hamster. Journal of Pineal Research, March, 1–15. https://doi.org/10.1111/jpi.12869

Sahu, A., & Verma, R. (2023). Bisphenol S dysregulates thyroid hormone homeostasis; Testicular survival, redox and metabolic status: Ameliorative actions of melatonin. Environmental Toxicology and Pharmacology, 104, 104300.

ThermoFisher Scientific. (n.d.). Real-Time PCR Hand-Book.

Wilson, K., & Walker, J. (2010). Principles and Techniques of Biochemistry and Molecular Biology (K. Wilson & J. Walker (eds.); Seventh). Cambridge University Press.

8

Estimation of Biochemical Parameters in Insects

Parinita Singh, Priyanka Yadav, Arvind Kumar Patel Sourabh Verma, Tamal Das, Pradeep Kumar, Simrata Dalpat Preeti Chhanda Sahoo and Bhupendra Kumar

Department of Zoology, Banaras Hindu University, Varanasi-221005, Uttar Pradesh, India

1. Introduction

Insects, like many other organisms, rely heavily on their diet for essential nutrients that regulate their growth and development. The quality of food intake significantly impacts various aspects of an insect's life cycle. Research, such as that conducted by Lee (2015) highlights that consuming a diet lacking in essential nutrients can lead to premature aging, a process known as senescence, in insects. Furthermore, such inadequate nutrition can hasten the rate of mortality as the insect ages. This underscores the critical role that proper nutrition plays in the longevity and overall health of insect populations. Moreover, studies conducted by researchers like Simpson et al. (2004), Behmer (2009), and Lee (2015) emphasize the importance of the ratio of dietary components, including proteins, carbohydrates, and lipids, in determining the fitness levels of various insect species. The balance and proportion of these nutrients in an insect's diet can significantly influence its ability to thrive, reproduce, and adapt to its environment. Understanding these dietary requirements and their effects on insect fitness is crucial for both ecological research and practical pest management strategies.

In the present chapter, we are providing the protocols to estimate the assimilation of glucose, soluble proteins, and triglycerides. These protocols are essential for understanding how insects metabolize and utilize key nutrients, shedding light on their physiological processes and energy requirements.

2. Estimation of glucose

Glucose quantification in whole body homogenates is carried out using a coupled colorimetric assay (Tennessen *et al.*, 2014). The insect are homogenized in a 1.5 ml centrifuge tube containing 200 µl of cold phosphate-buffered saline (PBS). Homogenates are centrifuged in a cooling centrifuge machine at 12,000 rpm for two minutes. The supernatant from each sample is collected and kept in a water bath at 70 °C for 10 minutes and then centrifuged again at 12,000 rpm for next two minutes. The supernatant from each sample is transferred to a fresh centrifuge tube, and the samples are stored at −20 °C in a mini cooler for later analysis.

2.1. Preparation of glucose standard

Glucose standards are prepared by diluting a 1mg/ml glucose standard with PBS to concentrations of 0.16 mg/ml, 0.08 mg/ml, 0.04 mg/ml, 0.02 mg/ml, and 0.01 mg/ml. Using a micro-pipette, 30 µl of the supernatant from each sample is transferred to wells in a flat bottom 96-well microplate. Additionally, 30 µl of glucose standards and 30 µl of PBS are transferred to two separate wells of the microplate to serve as the standard and blank, respectively. To each well containing the sample (standard solution or blank), 100 µl of glucose reagent is added. Microplates are sealed with aluminium foil to prevent evaporation, and the plates are incubated for next 15 minutes at room temperature. The total absorbance at 340 nm is measured using the microplate reader. Standard curves are prepared by plotting net absorbance *versus* the glucose concentration of each glucose standard. The total glucose concentration in each sample is determined by comparing the unknown samples to the prepared standard curves.

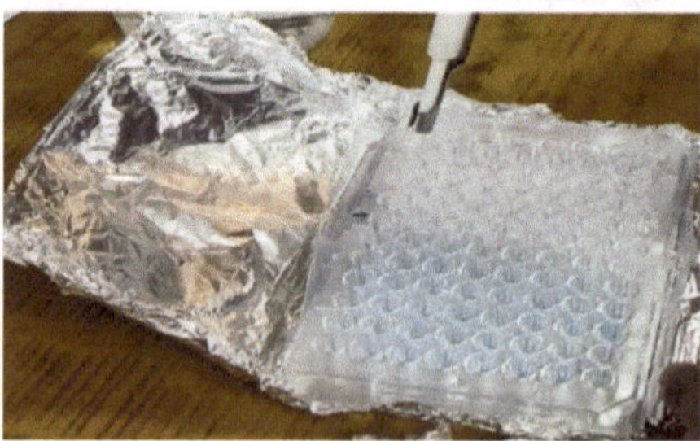
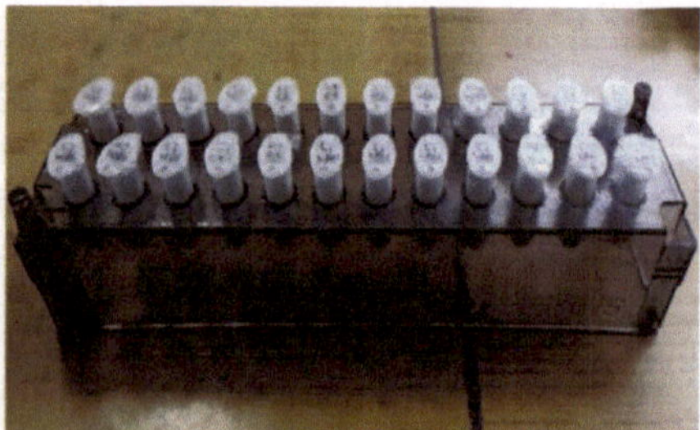

Fig. 1: Sample preparation for biochemical estimation

3. Estimation of proteins

Protein quantification in the whole body homogenates is performed using a Bradford protein assay (Bradford, 1976). Bradford reagent is prepared using Coomassie Brilliant Blue G-250 and orthophosphoric acid. Using a homogenizer, insects are homogenized in 1.5 ml centrifuge tubes containing 200 µl of cold PBS. Homogenates are then centrifuged in a centrifuge machine at 12,000 rpm for two minutes. The supernatant of each sample is collected in fresh centrifuge tubes and immediately kept in a mini cooler at −20 °C for further use.

3.1. Preparation of protein standard

Protein standards are prepared by serially diluting a 1 mg/ml Bovine Serum Albumin (BSA) protein standard.

Preparation of sample solution: Sample solutions are prepared by adding 1 µl of the respective sample to 9 µl of distilled water in a centrifuge tube. Then, 10 µl of the respective sample solution is transferred to a flat bottom 96-well microplate. Serially diluted BSA protein standards are also transferred to the microplate. Additionally, 10 µl of distilled water is transferred into two wells of the microplate as blank solutions. Next, 200 µl of Bradford reagent is added to each well containing the sample solution, standard solution, and blank solution, respectively, and mixed well. Microplates are sealed with aluminium foil to prevent evaporation and incubated for 8 minutes at 25 °C. The total absorbance at 595 nm is measured using the microplate reader. Standard curves are prepared by plotting net absorbance *versus* the protein concentration of each BSA protein standard. The total protein concentration in each sample is determined by comparing the unknown samples to the prepared standard curves.

4. Estimation of triglycerides

Lipid quantification in whole-body homogenates is conducted using a coupled colorimetric assay (Tennessen *et al.*, 2014). Insects are homogenized in 1.5 ml centrifuge tubes containing 200 µl of cold PBS + 0.05% Tween-20 (PBST). The homogenates are centrifuged at 12,000 rpm for two minutes. The supernatant is collected and kept in a water bath at 70°C for 10 minutes before being immediately transferred to a −20°C mini cooler. From the supernatant, 2 µl of the respective sample is transferred to a flat bottom 96-well microplate using a micro-pipette. Additionally, 2 µl of standard solution (triglycerides standard 200 mg/dl) is transferred to two wells of each microplate. To each well containing a sample or the standard solution, 200 µl of triglyceride enzyme reagent (Beacon; S13G) is added. The same volume of reagent is also added to two empty wells of the plate to serve as blanks.

The microplates are sealed with aluminium foil to prevent evaporation and incubated for 10 minutes at 37°C. Total absorbance at 490 nm is measured using a microplate reader, and the total triacylglycerol concentration is determined for each sample by subtracting the absorbance value of the blank well from the absorbance value of the sample well. The TAG content in each sample is calculated as:

$$\text{Concentration of Triglycerides (mg / dl)} = \frac{\text{Absorbance of sample}}{\text{Absorbance of standards}} \times 200$$

Conclusion

By accurately measuring the assimilation rates of glucose, soluble proteins, and triglycerides, researchers can gain insights into the efficiency of nutrient uptake and utilization in insect populations.

Acknowledgements

The authors are thankful to UGC-CAS and DST-FIST-funded Department of Zoology, Banaras Hindu University, Varanasi, India. The authors are also thankful to IoE, Banaras Hindu University, Varanasi for Incentive (Phase-IV) and Transdisciplinary Grants.

References

Behmer, ST (2009) Insect herbivore nutrient regulation. Annu Rev Entomol 54:165-187

Bradford MM (1976) A rapid and sensitive method for the quantitation of microgram quantities of protein utilizing the principle of protein-dye binding. Anal Biochem 72:248-254

Lee KP (2015) Dietary protein: carbohydrate balance is a critical modulator of lifespan and reproduction in *Drosophila melanogaster*: a test using a chemically defined diet. J Insect Physiol 75:12-19

Simpson SJ, Sibly RM, Lee KP, Behmer ST, Raubenheimer D (2004) Optimal foraging when regulating intake of multiple nutrients. Anim Behav 68:1299-1311

Tennessen JM, Barry WE, Cox J, Thummel CS (2014) Methods for studying metabolism in *Drosophila*. Methods 68:105-115

9

Determination of Enzymatic Antioxidants in Insects

Parinita Singh, Priyanka Yadav, Arvind Kumar Patel SourabhVerma, Tamal Das, Sikta Saroja, Paritosh Sharma Devendra Kumar Mishra and Bhupendra Kumar

Department of Zoology, Banaras Hindu University, Varanasi-221005, Uttar Pradesh, India

1. Introduction

Organisms are known to experience stress when environmental conditions, such as temperature, photoperiod, and humidity deviate from their optimal range. Organisms that respire aerobically constantly generate reactive oxygen species (ROS) through metabolism and employ antioxidants to reduce excess ROS formation and maintain a redox balance (Juan *et al.*, 2021). ROS play crucial roles in cell signaling (Sies and Jones, 2020), homeostasis, cell death, immune defense against pathogens, and the induction of mitogenic responses (Dröge, 2002). Therefore, optimal environmental conditions maintain a balance between ROS formation and antioxidant processes in all living organisms.

Adverse environmental conditions, whether lower or higher than optimal, can promote oxidative damage within the organism's body, leading to oxidative stress and ultimately resulting in the overproduction of ROS (Storey and Storey, 2012). In response to macromolecular damage, stress activates a universal defense mechanism known as the cellular stress response (Kültz, 2004). Insects, being the most common and diverse group of terrestrial organisms, tend to exhibit seasonality, which narrows the range of ambient temperature or humidity they can tolerate during their most active stages (Kishimoto-Yamada and Itioka, 2015). They also heavily rely on behavioural and physiological systems to regulate their body temperature, allowing them to survive in both hot and cold conditions.

While environmental stress can reduce the antioxidant status and may induce oxidative stress (An and Choi, 2010), the body protects itself from ROS damage during oxidative stress by producing enzymes such as superoxide dismutase (SOD), catalase (CAT), lactoperoxidase, glutathione peroxidase, and peroxiredoxin (Michiels et al., 1994). The primary antioxidant enzymes involved in ROS neutralization are SOD and CAT (Krishnan et al., 2007). Under environmental stress, SOD serves as the first line of defense against ROS-induced damage, catalyzing the conversion of superoxide into oxygen and hydrogen peroxide ($2O_2^- + 2H^+ \rightarrow H_2O_2 + O_2$) (Park et al., 2012), while CAT degrades H_2O_2 to H_2O and O_2 (Cui et al., 2023). Furthermore, ROS can initiate lipid peroxidation (LPO), targeting polyunsaturated fatty acids in lipids and disrupting membrane fluidity (Paital et al., 2016). Therefore, an elevated level of LPO serves as an indicator of oxidative stress.

The primary oxidation products of peroxidized polyunsaturated fatty acids are LPO aldehydes, such as malondialdehyde (MDA), which are used to assess the extent of LPO and act as biological markers of oxidative stress (Zeis et al., 2019). Oxidative stress can ultimately lead to oxidative damage to DNA (Poetsch, 2020), lipids and proteins (Stadtman, 2006), and significantly impacting the lifespan of insects (Ristow and Schmeisser, 2011). In the present chapter we will discuss the protocols to quantify the antioxidant enzyme activity in insects. By measuring the activity levels of enzymes such as SOD, CAT and LPO products, researchers can gain insights into an insect's ability to combat oxidative stress.

2. SOD activity assay

The larvae and adults are homogenized separately (one individual at a time) in 1.5 ml centrifuge tubes in ice-cold Phosphate Buffer Saline (PBS) (pH 7.0) at 10% (w/v) ratio using a tissue homogenizer for 30 seconds at 4°C. Homogenates are centrifuged at 10,000 g for 10 minutes at 4°C in a centrifuge. The supernatant collected is used as an enzyme source. In a fresh centrifuge tube, 700 µL of the reaction buffer is prepared using PBS=550 µL, L-methionine = 40 µL (20 mM), Triton X-100 = 20 µL (1%), Hydroxylamine-hydrochloride = 40 µL (10 mM), and Ethylenediaminetetraacetic acid (EDTA) = 50 µL (50 mM). Within the reaction buffer, 50 µL of the sample is added, whereas in reference and blank solutions, 50 µL PBS is added.

The above tubes are transferred to an incubator for pre-incubation at 37°C for 5 minutes. Thereafter, 40 µL (50 µM) of Riboflavin is added to both the sample and the reference solutions. Then the reaction mixture tubes containing the samples are exposed to 15 W CFL light for 10 minutes and subsequently 500 µL of freshly prepared Greiss reagent is added to all the tubes. The UV-visible spectrophotometer is set up at 543 nm and calibrated with the blank solution, and the absorbance of the sample and reference is recorded. The SOD activity calculation is done using the following formula: SOD (U/mg) = {(V_0/V)-1/ mg of protein}, where V_0 is the absorbance of the reference solution and V is the absorbance of the sample solution. The enzyme activity is expressed as U/mg protein. One unit of SOD activity is defined as the amount of enzyme capable of inhibiting 50% of nitrite formation under assay conditions (Das *et al.*, 2000).

The protein concentration of the enzyme is calculated by the method from Bradford (1976) using BSA as standard; and the total protein concentration in each sample is determined by comparing the unknown samples with the prepared standard curve. All readings are recorded in triplicate. The average of three readings is considered as a single value for statistical analysis.

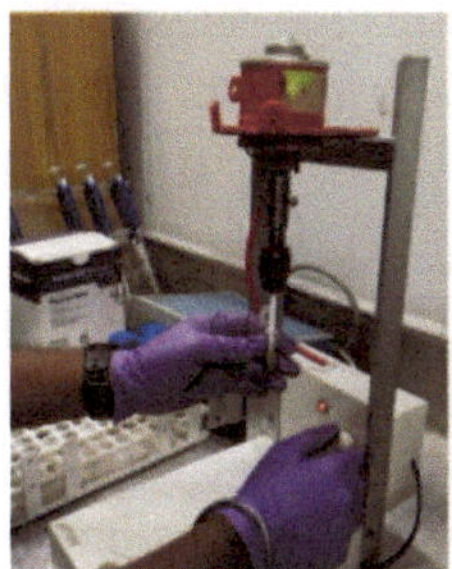

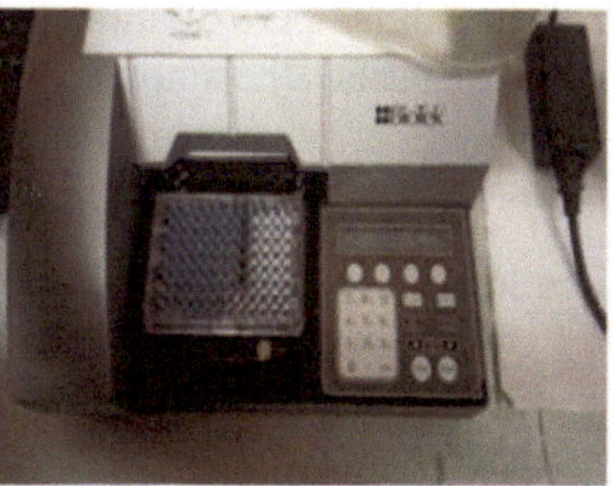

Fig. 1: Sample preparation for enzymatic assay

3. CAT activity assay

To determine the CAT activity (see Aebi, 1984), the enzyme is extracted as per the protocol described in the SOD assay above. The supernatants containing the enzyme extracts are collected in new centrifuge tubes and used for enzymatic assay. 10 μL of the enzyme extract diluted with PBS (690 μL) is added to a centrifuge tube containing 300 μL of 30 mM H_2O_2 and the tube is quickly placed in a UV-visible spectrophotometer kept at 240 nm to record the absorbance with a recorder at 30 second intervals for 3 minutes. The readings are recorded in triplicate and the average of the three readings is considered as a single value for further statistical analysis. Calculations of CAT activity are done with the help of the following formula:

$$\text{CAT activity (U / mg)} = \frac{(A_0 - A_{180}) \times V_t}{\varepsilon_{240} \times d \times V_s \times C_t \times 0.001}$$

where (A_0 - A_{180}) is the difference between the initial and final absorbance; V_t is the total volume of the reaction (1 mL); ε_{240} is the molar extinction coefficient for H_2O_2 at OD_{240} (34.9 mol^{-1} cm^{-1}); d is the optical path length of the cuvette (1 cm); V_s is the volume of the sample (10 μL); C_t is the total protein concentration in the sample; and 0.001 is the absorbance change caused by 1 U of enzyme per minute at 240 nm OD. The protein concentration is calculated by the method from Bradford (1976) using Bovine Serum Albumin (BSA) (SRL, India) as the standard, and the total protein concentration in each sample is determined by comparing the unknown samples to the prepared standard curve.

4. Extraction and estimation of thiobarbituric acid-reactive substances (TBARS) content (LPO assay)

The TBARS assay (Ohkawa *et al.*, 1979) is a common method used to measure lipid peroxidation, which is an indicator of oxidative stress. The assay quantifies the amount of malondialdehyde (MDA), a by-product of lipid peroxidation, present in the sample. The Thiobarbituric Acid (TBA) reacts with MDA, forming a pink-coloured complex.

The larvae and adults are homogenized separately in PBS (10% w/v) followed by centrifugation at 3,000 g for 10 minutes at 4°C and the supernatant is collected in a fresh centrifuge tube. 50 μL of the supernatant from each sample is added to 825 μL of thiobarbituric acid (TBA) reagent prepared using 50 μL (8%) of sodium dodecyl sulfate (SDS), 375 μL (20%) of acetic acid, 375 μL (0.8%) of an aqueous solution of thiobarbituric acid, and 25 μL (0.8%) of butylated hydroxy toluene. The reaction mixture containing the sample is incubated at 95°C in a water bath, which resulted in the appearance of a pinkish colour. After 1 hour, the tubes are taken out, cooled, and centrifuged at 3000 g for 10 minutes to remove any debris. The supernatant (800 μl) is collected and the absorbance is measured at 532 nm in a UV-visible spectrophotometer. The amount of TBARS present is calculated using the following formula:

$$\text{nmol TBARS / mg protein} = \frac{V \times OD}{156 \times C}$$

Where V is the reaction mixture volume (μl); OD is the absorbance at 532 nm; 156 is the extinction coefficient of the MDA-TBA complex at 532 nm ($mmol^{-1}$ cm^{-1}); C is the protein concentration (mg). The LPO activity is expressed as nmol of TBARS produced per mg protein. The protein concentration of the enzyme is calculated by the method from Bradford (1976) using BSA as standard; and the total protein concentration in each sample is determined by comparing the unknown samples with the prepared standard curve.

Conclusions

Studying the antioxidant enzyme activity provides invaluable insights into the intricate mechanisms by which antioxidants uphold cellular homeostasis and counteract oxidative damage. By meticulously examining the activity levels of key antioxidant enzymes such as superoxide dismutase (SOD) and catalase (CAT), researchers can uncover the dynamic interplay between these enzymatic defenses and reactive oxygen species (ROS) within biological systems. Through this investigation, researchers can discern how antioxidant enzymes effectively scavenge ROS, thereby safeguarding cellular components such as lipids, proteins, and DNA from oxidative harm. Furthermore, understanding the nuanced regulation of antioxidant enzyme activity sheds light on the adaptive responses of organisms to various environmental stressors, including exposure to pollutants, fluctuations in temperature, and pathogenic assaults.

Acknowledgements

The authors are thankful to UGC-CAS and DST-FIST-funded Department of Zoology, Banaras Hindu University, Varanasi, India. The authors are also thankful to IoE, Banaras Hindu University, Varanasi for Incentive (Phase-IV) and Transdisciplinary Grants.

References

Aebi, H., 1984. Catalase in vitro. Meth. Enzymol. 105, 121–126.

An, M.I., Choi, C.Y., 2010. The activity of antioxidant enzymes and physiological responses in ark shell, *Scapharca broughtonii*, exposed to thermal and osmotic stress: Effects on hemolymph and biochemical parameters. Comp. Biochem. Physiol. B, Biochem. Mol. Biol. 155, 34–42.

Bradford, M.M., 1976. A rapid and sensitive method for the quantitation of microgram quantities of protein utilizing the principle of protein-dye binding. Anal. Biochem. 72, 248–254.

Cui, Z., He, F., Li, X., Li, Y., Huo, C., Wang, H., Qi, Y., Tian, G., Zong, W., Liu, R., 2023. Response pathways of superoxide dismutase and catalase under the regulation of triclocarban-triggered oxidative stress in *Eisenia foetida*: Comprehensive mechanism analysis based on cytotoxicity and binding model. Sci. Total Environ. 854, 158821.

Das, K., Samanta, L., Chainy, G., 2000. A modified spectrophotometric assay of superoxide dismutase using nitrite formation by superoxide radicals. Indian J. Biochem. Biophys. 37, 201–204.

Dröge, W., 2002. Free Radicals in the Physiological Control of Cell Function. Physiol. Rev. 82, 47–95.

Juan, C. A., Pérez de la Lastra, J.M., Plou, F.J., Pérez-Lebeña, E., 2021. The Chemistry of Reactive Oxygen Species (ROS) Revisited: Outlining Their Role in Biological Macromolecules (DNA, Lipids and Proteins) and Induced Pathologies. Int. J. Mol. Sci. 22, 9.

Kishimoto-Yamada, K., Itioka, T., 2015. How much have we learned about seasonality in tropical insect abundance since Wolda (1988)?. Entomol. Sci. 18, 407-419.

Krishnan, N., Kodrík, D., Turanli, F., Sehnal, F., 2007. Stage-specific distribution of oxidative radicals and antioxidant enzymes in the midgut of *Leptinotarsa decemlineata*. J. Insect Physiol. 53, 67–74.

Kültz, D., 2004. Molecular and evolutionary basis of the cellular stress response. Annu. Rev. Physiol. 67, 225–257.

Michiels, C., Raes, M., Toussaint, O., Remacle, J., 1994. Importance of SE-glutathione peroxidase, catalase, and CU/ZN-SOD for cell survival against oxidative stress. Free Radic. Biol. Med. 17, 235–248.

Ohkawa, H., Ohishi, N., Yagi, K., 1979. Assay for lipid peroxides in animal tissues by thiobarbituric acid reaction. Anal. Biochem. 95, 351–358.

Paital, B., Panda, S.K., Hati, A.K., Mohanty, B., Mohapatra, M.K., Kanungo, S., Chainy, G.B.N., 2016. Longevity of animals under reactive oxygen species stress and diseasesusceptibility due to global warming. World J. Biol. Chem. 7, 110–127.

Park, S.Y., Nair, P.M.G., Choi, J., 2012. Characterization and expression of superoxide dismutase genes in *Chironomus riparius* (Diptera, Chironomidae) larvae as a potential biomarker of ecotoxicity. Comp. Biochem. Physiol. C Toxicol. Pharmacol. 156, 187-194.

Poetsch, A.R., 2020. The genomics of oxidative DNA damage, repair, and resulting mutagenesis. Comput. Struct. Biotechnol. J. 18, 207–219.

Ristow, M., Schmeisser, S., 2011. Extending life span by increasing oxidative stress. Free Radic. Biol. Med. 51, 327–336.

Sies, H., Jones, D.P., 2020. Reactive oxygen species (ROS) as pleiotropic physiological signalling agents. Nat. Rev. Mol. Cell Biol. 21, 363–383.

Stadtman, E.R., 2006. Protein oxidation and aging. Free Radic. Res. 40, 1250–1258.

Storey, K.B., Storey, J.M., 2012. Insect cold hardiness: metabolic, gene, and protein adaptation. Can. J. Zool. 90, 456–475.

Zeis, B., Buchen, I., Wacker, A., Martin-Creuzburg, D., 2019. Temperature-induced changes in body lipid composition affect vulnerability to oxidative stress in *Daphnia magna*. Comp. Biochem. Physiol. B, Biochem. Mol. Biol. 232, 101–107.

Index

Index